Geographies of Making, Craft and Creativity

This book brings together cutting-edge research from leading international scholars to explore the geographies of making and craft. It traces the geographies of making practices from the body, to the workshop and studio, to the wider socio-cultural, economic, political, institutional and historical contexts. In doing so it considers how these geographies of making are in and of themselves part of the making of geographies. As such, contributions examine how making bodies and their intersections with matter come to shape subjects, create communities, evolve knowledge and make worlds.

This book offers a forum to consider future directions for the field of geographies of making, craft and creativity. It will be of great interest to creative and cultural geographers, as well as those studying the arts, culture and sociology.

Laura Price is a feminist cultural geographer. Her PhD thesis explored the geographies of knitting, gendered creativity and the role of materials in everyday lives. She is currently a writer of education resources at the Royal Geographical Society (with the Institute of British Geographers) where she works with universities, schools and teachers to promote and share geographical learning. She is co-editor of *Geographies of Comfort* (Routledge, 2018).

Harriet Hawkins is based in the Department of Geography, Royal Holloway, University of London where she works on the geography of art works and art worlds. She is committed to practice-based research and collaboration with artists and arts institutions and organisations. She is author of *For Creative Geographies* (Routledge, 2013) and *Creativity: Live, Work, Create* (Routledge, 2016), editor of *cultural geographies* and founder and co-Director of Royal Holloway, Centre for the GeoHumanities, where she is currently Professor of GeoHumanities.

Routledge Research in Culture, Space and Identity

Series editor: Dr. Jon Anderson, School of Planning and Geography, Cardiff University, UK

The *Routledge Research in Culture, Space and Identity Series* offers a forum for original and innovative research within cultural geography and connected fields. Titles within the series are empirically and theoretically informed and explore a range of dynamic and captivating topics. This series provides a forum for cutting edge research and new theoretical perspectives that reflect the wealth of research currently being undertaken. This series is aimed at upper-level undergraduates, research students and academics, appealing to geographers as well as the broader social sciences, arts and humanities.

For a full list of titles in this series, please visit www.routledge.com/Routledge-Research-in-Culture-Space-and-Identity/book-series/CSI

Geographies of Making, Craft and Creativity

Edited by
Laura Price and Harriet Hawkins

Routledge
Taylor & Francis Group

LONDON AND NEW YORK

First published 2018
by Routledge

2 Park Square, Milton Park, Abingdon, Oxfordshire OX14 4RN
52 Vanderbilt Avenue, New York, NY 10017

Routledge is an imprint of the Taylor & Francis Group, an informa business

First issued in paperback 2020

British Library Cataloguing in Publication Data
A catalogue record for this book is available from the British Library

Library of Congress Cataloging in Publication Data
A catalog record for this book has been requested

ISBN: 978-1-138-23874-9 (hbk)
ISBN: 978-0-367-59170-0 (pbk)

Typeset in Times New Roman
by Taylor & Francis Books

We dedicate this book to Phil Crang – an inspirational and generous supervisor and mentor – thank you.

Contents

Figures

Contributors

Miriam Burke is a practicing artist and cultural geography PhD student at Royal Holloway, University of London. Her research practice involves working with community groups exploring environmental issues through participatory art as well as collaborating with scientists. She has worked as an arts and environment consultant for Manor House Development Trust and co-designed and taught on the 'Inspirational Landscapes' module at Keele University, which was awarded an innovation in teaching award.

Chantel Carr is an Associate Research Fellow and PhD Candidate at the University of Wollongong, Australia. Her research investigates how material work is being impacted by environmental and economic change. Chantel has undertaken ethnographic fieldwork with industrial repair and maintenance workers, architect-builders, users of maker-spaces, and recently commercial building managers. Prior to her PhD Chantel trained and worked in architecture and design.

Rebecca Collins is Senior Lecturer in Human Geography at the University of Chester. Her research interest in making and mending grew out of doctoral research into the sustainability implications of the material culture of youth. Her more recent projects have explored mending culture in maker-spaces, the notion and practice of 'enough' in everyday youth cultures, and the aesthetics of product ageing.

Zoe Collins graduated in BA Geography at Royal Holloway, University of London in 2016, and in 2017 she completed a Masters' in Environment, Development and Policy at the University of Sussex. Her work lies at the intersection between creativity, gender, and development. She has a particular interest in labour rights and supply chain politics.

Dydia DeLyser is a cultural-historical geographer in the Department of Geography & the Environment at California State University, Fullerton. She has interests drawing together qualitative, participatory research, embodied practices, and historical artifacts. Her research typically focuses on Los Angeles, and she is at work on a book about the historical geography of the neon-sign industry in that city, and in the USA as a whole.

Caitlin DeSilvey is Associate Professor of Cultural Geography at the University of Exeter, where she has been employed since 2007. Her research explores the cultural significance of material and environmental change, with a particular focus on heritage contexts. Recent publications include *Anticipatory History* (Uniformbooks, 2011, with Simon Naylor and Colin Sackett), *Visible Mending* (Uniformbooks, 2013, with Steven Bond and James R. Ryan) and *Curated Decay: Heritage Beyond Saving* (University of Minnesota Press, 2017).

Tim Edensor teaches Cultural Geography in the Department of Geography Manchester Metropolitan University. He is the author of *Tourists at the Taj* (Routledge, 1998), *National Identity, Popular Culture and Everyday Life* (Berg, 2002) and *Industrial Ruins: Space, Aesthetics and Materiality* (Berg, 2005), *From Light to Dark: Daylight, Illumination and Gloom* (Minnesota, 2017) and the editor of *Geographies of Rhythm* (2010). Tim has written extensively on national identity, tourism, ruins and urban materiality, mobilities and landscapes of illumination and darkness.

Chris Gibson is Professor of Human Geography and Director of the interdisciplinary research programme, Global Challenges: Transforming Lives & Regions, at the University of Wollongong, Australia. He has written on diverse topics – from the social and economic contributions of festivals to regional communities, to the history and geography of niche industries including music, surfboard-making and guitar manufacturing. He is currently Editor-in-Chief of the academic journal, *Australian Geographer*.

Helen Holmes is a Hallsworth Research Fellow in Sociology at the University of Manchester. Her work centres upon the juncture of temporality, materiality and practice to explore the everyday. Her current work focuses on exploring contemporary forms of thrift and notions of make do and mend through the project: www.makersmakedoandmend.org. Helen has published in a variety of journals including: *Work, Employment & Society, Time & Society* and *Economy & Society*.

Richard E. Ocejo is associate professor of sociology at John Jay College and the Graduate Center of the City University of New York (CUNY). He is the author of *Masters of Craft: Old Jobs in the New Urban Economy* (Princeton University Press, 2017), and of *Upscaling Downtown: From Bowery Saloons to Cocktail Bars in New York City* (Princeton University Press, 2014). His work has appeared in such journals as *City & Community, Poetics, Ethnography,* and the *European Journal of Cultural Studies*.

James R. Ryan is Head of Programme for the MA History of Design–Material Culture, Performance, Photography at the V&A Museum, London, and Associate Professor in Historical and Cultural Geography, University of Exeter. Recent publications include *Photography and Exploration* (Reaktion Books, 2013); *Visible Mending* (Uniformbooks, 2013 with Steven

Bond and Caitlin DeSilvey); and *New Spaces of Exploration: Geographies of Discovery in the Twentieth Century* (IB Tauris, 2010, edited with Simon Naylor).

Jenny Sjöholm is a lecturer at Linköping University, Sweden; and a Marie Sklowdowska–Curie Research Fellow in GeoHumanities at Royal Holloway, University of London (2017–2018). Her research interests concern the geographies and politics of contemporary art, artistic practice and cultural work. Currently she is working on an interdisciplinary project on the construction of value in the contemporary art world with a specific focus on privatization and the role of private art collectors.

Elyse Stanes is a PhD Candidate in the School of Geography and Sustainable Communities at the University of Wollongong, Australia. Her primary research interests include young people, consumption and questions of sustainability. Elyse's PhD research explores cultural-material cultures of fashion consumption among young people – with a particular focus on the material and practice of wearing clothes.

Elizabeth R. Straughan is currently working in Teaching Support at the University of Melbourne. Her research unravels the volatile nature of the body through attendance to the skin as well as the material and metaphorical dynamics of touch. She has published in international journals such as *Cultural Geography, Emotion, Space and Society* and *Geography Compass*.

Nicola Thomas is Associate Professor in Cultural and Historical Geography at the University of Exeter. She has developed a body of work around craft geographies, situating contemporary and 20th century craft practice within the broader creative economy. Her approach addresses the intersection of material, historical, cultural, social, political and economic contexts through an exploration of craft makers' livelihoods and the spatial dimension of their labour. Her research always attends to the historicity of cultural production and consumption, bringing a historical sensitivity to critical understandings of the cultural and creative economy.

Andrew Warren is Lecturer in Economic Geography at the University of Wollongong, Australia. His research focuses on labour and the shifting geographies and cultures of work, and impacts of industrial and technological change. Empirically his research focuses on industrial cities and regions, and geographies of manufacturing. He is the author of *Surfing Places, Surfboard Makers* (University of Hawaii Press, 2014).

Acknowledgements

This book has been slow in the making, perhaps fittingly. It has its formal origins in a series of sessions organized at the annual meeting of the Association of American Geographies in Tampa and the Royal Geographical Society Annual Conference in London, both in 2015.

We wish to thank all the authors and our editors at Routledge in particular Faye Leerink and Ruth Anderson for their patience with the slow evolution of this book, we think, we hope, it is better for the time.

Our interests in the geographies of making date further back however and encompass discussions and support from a range of colleagues and friends. For both of us an interest in making stems from our families and the materials that have made our lives familiar. Whether it be knitting, carding and spinning wool or the making of everything from sheds to airplanes from stacks of wood. It stems from the valuing of the vernacular and the everyday. In some ways, it is easy to research craft and making; hand-made objects inspire storytelling. Who made it? For whom? Why has it been kept, or mended or broken? Craft and making tells us stories of everyday lives and we feel thankful for those who have shared their stories along the way.

For Harriet, this life experience evolved into an intellectual interest through her doctoral work at Nottingham on artists working with rubbish and waste, under the supervision of Stephen Daniels and Nick Alfrey, her postdoctoral work with David C. Harvey and Nicola Thomas at University of Exeter, and with Deborah Dixon and Libby Straughan at Aberystwyth University and Sallie Marston and John Paul Jones III at University of Arizona. Whilst for Laura, her studies of floristry at University of Manchester and of knitting, under the supervision of Phil Crang, at Royal Holloway, University of London, underpin her interest in geographies of making.

For Laura, this intellectual work began whilst studying gender, creativity and embroidered mapping at undergraduate level under the supervision of Chris Perkins at the University of Manchester. An interest in material lives and the power of material at the University of Sheffield under the supervision of Matt Watson led to doctoral research into knitting, wool and textile art at the Department of Geography, Royal Holloway, University of London, under the supervision of Phil Crang, with Harriet Hawkins as advisor.

As ever the production of books takes a village. In addition to the authors of the chapters in this text, we both want to thank staff and students at Royal Holloway, University of London, especially the Landscape Surgery group (particularly Miriam Burke, Ella Harris, Flora Parrott, Danny McNally, Sofie Narbed) and Phil Crang, as well as Chris Gibson at the University of Wollongong and Dydia DeLyser, Cal State Fullerton and Mark Jayne, Cardiff University for their discussion and support of ideas contained within the text.

For the kind of support and friendship that makes writing books fun and engaging and for reminding her, often through their own diverse practices, what making is about, Harriet would like to thank all the artists and makers she has worked with over the years, but especially Katherine Brickell, Mim Burke, Luce Choules, Tania El Khoury, Anja Kanngieser, Flora Parrott, Libby Straughan and of course Anne, Louis and Joseph Hawkins. She would also like to thank the Arts and Humanities Research Council for their support of some of the research and teaching relief that went into the production of this volume: AH/N004132/1.

Laura would like to thank each knitter, maker and artist she has worked with during her research – she has learnt much about generosity, humour and the time it takes to finish projects well (or sometimes, to just finish). Whilst a memorably terrible knitter herself, she thanks Rose and Mavis for their perseverance and love. As ever, she would like to thank Martin, Mandy, Ryan and Kieran for their support. Both Harriet and Laura would like to thank Flora Parrott for the use of her image for the front cover of the book.

1 Towards the geographies of making

An introduction

Harriet Hawkins and Laura Price

Sheds, streets, kitchens, gardens, workshops, factories, studios, hair salons, maker spaces, butchers, barbers, cocktail bars, craft guilds, community centres and bodies (human and non-human), such is the diverse patchwork of sites at which making takes place – in this book alone. Making appears to be firmly part of the zeitgeist. It resonates with academic concerns with embodiment, practice and matter and their intersections with our ways of knowing and being in the world (Anderson and Harrison, 2010; Bennett, 2012) as well as capturing the imaginations of the public, economists, trend-setters and politicians alike, with fashions for making, crafting and tinkering well documented in a host of popular books (Sennett, 2008; Crawford, 2010; Gauntlett, 2011; Anderson, 2013; Hitch, 2013; Lang, 2013; Heatherwick and Rowe, 2015; Korn 2017). In the UK, television programmes like 'The Great Sewing Bee' or 'The Great Pottery Throw Down' occupy prime-time viewing slots. While American magazines such as *The Atlantic* and *The New Yorker* feature accounts of the maker movement sweeping US rust belt cities and Chinese technopolises alike. Such articles recount a revolution in production occurring via a revaluation of light industrial skills that are practiced in collective 'maker' spaces kitted out with everything from welding torches to 3D printers (Morozov, 2014; Fallows, 2016). Glossy lifestyle magazines and colour supplements read on lazy sunny Sundays and during precious downtime in bustling hairdressers around the world frame handicrafts as the latest *trend-du-jour* for aspirational hipster urban elites. Whilst, for those seeking material and mental solace from busy consumer worlds, the mindful and environmentally aware ethics of the slow movement provides the time, space and aesthetic context within which to revalue the handmade, and practices of repair and restoration. Yet despite the prominence of making in the early decades of the twenty first century, we should heed those warnings issued around the knitting trend, that "despite the cries of the press that 'knitting is back!' it in fact never went away" (Dirix, 2014: 92). The contemporary rise in making takes place across a backdrop of long practiced traditions and skills that have for centuries been woven into the fabric of landscapes and lives. Indeed, this complex and rich history informs its popular appeal. Yet as Hall and Jayne's

(2015) study of dressmaking cautions us around these temporalities, contemporary cultures and geographies of dressmaking are, they argue, "distinguishable from their historic, post-war counterpart, and that to simply denote current dressmaking practices as being a mere replication of the 'make-do-and-mend' mentality is to overlook that austerity today comes from a very different set of social, economic and political conditions to those that have come before." However the intermesh of the historical and contemporary plays out, such collective concerns with making represents, we argue, a requirement that we revisit and re-negotiate the spaces and practices of the production of things, and that we interrogate the politics therein.

The visitation and revaluation of the spaces, practices and politics of making carried out in the 12 chapters collected in this volume is shaped by a concern with the geographies of making. To read scholarship on making is to become enmeshed in a diverse weave of threads that unravel from a variety of conceptual skeins. These include economic, political and social research on the creative economy and the historical and cultural discussions of the material, practiced and embodied dimensions of craft practices and vernacular creativities. We could also add to this list a growing sense of scholarship driven by environmental concerns that urge the Western world toward a re-engagement with vital materials and volatile matters in a world under threat. Despite its diversity, this work is linked by a strong sense that to attend to the geographies of making (i.e. where production takes place) is also to comprehend how geographies are themselves brought into being. This is, in short, the central premise that weaves across this collection, that to appreciate the geographies of making is at one and the same time to appreciate the making of geographies. Making emerges from the threads of empirical material and conceptual discussion as an embodied, material, relational and situated practice that spins connections between corporeal practices and formal institutional and political spaces, between governance and policy practices and practices of resistance, and between highly professionalised practices as well as amateur, vernacular and mundane practices. As such, the empirical fabric of these chapters textures our sense of how making shapes places and communities, transforms spaces and politics, and forges environmental relations.

In what follows we want to unfold the intersections of practices of making and the geographies being made. We follow this with a reflection on the methodological challenges of studying the geographies of making. Together these sections offer a synopsis of the volume and the content and intersections of its 12 chapters. We close the volume with a proposition of some new territories for the geographies of making. We reflect on the text's geographic and conceptual blind spots, and in doing so identify some of the threads we would like to see being picked up within any future agenda for the geographies of making.

Remaking our understandings of making

Making appears to be joining that host of 'key words' that are 'significant, binding words' like culture and nature which are recognised as hard to define (Williams, 2014: 15). Indeed, it has become something of a truism to preface discussions of making (and similarly craft) with observations concerning its definitional breadth but also its fundamental role in human 'being'. To ask 'what do we mean by making?' is to be greeted with the unfurling of sites and practices. Carr and Gibson (2015: 1) for example recount how "as humans we make bodies, homes, identities and memories every day. As a society we make landscapes, cities, decisions and structures for governing. And in daily work, the stuff that surrounds us is made." Given the complexities of asking "What do we mean by making?" this book does not directly seek an answer. Rather, gestures are assembled and ideas accumulate through the assonances and dissonances in the empirical and conceptual discussions within the chapters. Stitched together here are craft practices – whether contemporary, hipster or traditional – service labour and light and heavy industrial practices as well as vernacular, everyday making practices. Making negotiates its status as a labour practice, which has a politics like any other, as well as being an embodied sensory practice, a cerebral practice, a labour of love, a mundane practice and a routine practice of human and mechanic automation. There is a demand for an interweaving of the professional and amateur as well as the intersection of the production, consumption and circulation of objects and practices. There is also resistance to sorting making into practices undertaken within public or private worlds, or within urban and rural settings. Instead, making emerges as an agentive force that remakes such spatial distinctions. What evolves are understandings that eschew binary distinctions that work to unsettle spatial and social categories, in the place of a unification through a common focus on the critical force of making's geographies. To aid in thinking about these geographies we draw on the two strands of interdisciplinary scholarship indicated in the book's title; an extensive one around creativity, and a smaller more defined field concerned with craft.

Creativity, often thought of as the oil of the twenty-first century, has become a vast and diverse field of scholarship of late (see summary in Hawkins, 2016; Mould 2015). Indeed, creativity has become something of an imperative, full of promise, it has been variously situated as an economic saviour

> as a tool of neo-liberal politics and part of the diplomatic arsenal of state-craft practices, as a psychological trait and philosophical concept as well as an embodied, material and social practice that produces both highly specialist cultural goods and is a part of everyday life, holding within it myriad possibilities for making alternative worlds.
>
> (Hawkins, 2016: 1)

Much attention, within and beyond geography has been paid to the sites at which creativity takes place; currently cross-scalar explorations that encompass creative cities, clusters and networks, as well as discussions of making in the home or in the studio (Bain 2004; Edensor et al. 2009; Brace and Jones-Putra, 2010; Harvey et al., 2013; Sjoholm, 2014). Interestingly, within such work there is a growing concern with the nature and form of creative labour, for some akin to the aesthetic and embodied labour of the service economy (Ocejo, 2014b), for others renamed playbour or passion work after the associations of these practices with leisure and pleasure rather than the hard graft of a 'job' (Kucklich, 2005; Ardivisson et al., 2010). Of growing importance too is scholarship that explores creativity beyond the economic and political narratives that often foreground professionalised urban practices. This research pays attention to more everyday, community spirited, and mundane practices, often termed vernacular creativities (Burgess, 2006; Edensor et al., 2009). Occurring in overlooked spaces, these everyday practices not only expand what creativity is, but *where* it takes place, drawing our attention not only beyond the urban, but also to overlooked or marginalised spaces – from sheds, to homes, garages and suburbs.

One of the threads that can be followed throughout such diverse research on creativity is a concern with practice itself, its nature, as well as what such 'doings' achieve. This is perhaps most concentrated in that dispersed body of work – from across anthropology, sociology and geography, amongst other locations – that foregrounds the creative process, in contrast to studies of finished things and products. Leading the way conceptually is Ingold's anthropological discussion *Making* (2013), and a suite of papers and jointly edited collections that explore making as part of a wider set of creative processes alongside improvisation and growing (Hallam and Ingold, 2008, 2014). Here, making evolves as an interaction of sentient practitioners and active materials in the generation of form. Similarly concerned with addressing 'bodies that make' are ethnographic accounts of the production of highly crafted objects and their wider spatialities and socialities. We might think, for example, of Warren's and Gibson's (2014) account of the practices and cultures of surfboard making, Dudley's (2014) ethnography of artisanal guitar making, O'Connor's (2007) expansive research on glass blowing, or the growing body of work focused on knitting (Mann, 2015; Price, 2015). Such attentiveness to the 'doings' of making is both a product of, but also produces, the recovery of the body within and beyond geography, and the accompanying turns to materiality and practice (Anderson and Harrison, 2010; Bennett, 2012).

The second strand of work that this volume engages is scholarship on craft. The long and deep vein of literature on craft traditions and craft practices has been joined in the last decade or so by the rapid evolution of craft studies. This was solidified with the launch of the *Journal of Modern Craft* (2008) and *Craft Research* (2010), as well as the publication of the *Craft Reader* (2010). These intersect with the production of landmark texts such as Adamson's

Thinking Through Craft (2007) and *The Invention of Craft* (2014), together with volumes that explore the place of craft in contemporary art (Buzek, 2011), or that examine increasingly common practices such as yarn bombing (Moore and Prain, 2009). Often overlapping with scholarship on creativity, critical perspectives on craft address the distinct labour, bodily experiences, and social and material knowledge that define craft and that are shaped by, and actively shape relationships, communities and place (Ingold, 2013). As well as very traditional 'craft' practices (knitting, patchwork, pottery, woodwork and so on), increasingly the phrase has been evolved to recognise the forms of work being done in industries not usually defined as 'craft', such as cocktail making, or hairdressing (Sennett, 2008; Ocejo, 2014a; Holmes, 2015) and even research methods. Interestingly, much contemporary literature on craft has honed its critical edge precisely because craft has become so trendy, within and beyond the academy. Whilst there was long a sense of craft as being maligned, underappreciated, undervalued, or in some way 'in peril', this no longer seems to be the case (Adamson, 2007). As Jefferies observes (2011: 224), "craft has become the new cool, the new collectible: a rebellion against high-street branding and mall sameness alike, against the globalisation of labour exploitation and consumer indifference" (Jefferies, 2011: 224). Elizabeth Nathanson (2013: 109) however notes, "supposedly 'hip' craft artists distinguish themselves by first acknowledging the associations of crafts with a conservative aesthetic and 'elderly' point of view, and then by reinventing these crafts with a knowing and often ironic eye". Central to many of these debates are concerns with gender. Whilst Richard Sennett's (2008) urging of attention to 'craftsmanship' in contemporary society rather unfortunately failed to address issues of gender, feminist art historians have long explored the relationship between making and diverse identities (Pollock, 1999). Recently, for example, the prominence of hand-made objects such as the pink pussy hat worn at international protest marches in January 2017 has brought into popular discourse discussions around the power of craft and bodies that make.

These two brief sketches serve less to detail these overlapping strands of scholarship, than to stake out the ground across which our discussions of making take place. We are concerned to both stitch together these different threads of the discussions of making practices, but also to allow them to be unpicked, or unravelled in relation to one another. Such that what emerges are understandings of making that sometimes seem incongruent across time and space, but also across disciplinary and sub-disciplinary affiliations. The remainder of this section explores some of the key threads that weave across the chapters that follow. Taking in turn, and somewhat artificially, the geographies of making – sites and mobilities; skilled bodies, materialities and remakings; and then turning to the making of geographies, examining communities and places, as well as politics, subjects and environments. The discussion which follows focuses in on what kinds of questions and priorities emerge from these overlapping threads of discussion.

Unsettling sites of making

Where making takes place matters. Like the wider geographies of creativity, we might reflect upon the multiple, situated contexts of making. Bain (2004: 425), for example, describes arts practice as "embedded within the culturally constructed context of the art world and located within the place-based culture of the studio, the home, the neighbourhood, the community, the city, the nation". Geographies of creativity have, of late, been undergoing an important territorial reorientation, which demands that we expand our sense of those locations we consider to be sites of making. To take perhaps the clearest example of this reorientation, Gibson et al. (2012: 3) observe:

> Researchers have looked for creativity in fairly obvious places (big cities, cities making overt attempts to reinvent themselves through culture, creativity and cosmopolitanism); have found it there; and have theorised about cities, creative industries and urban transformation as if their subsequent models or logic were universally relevant everywhere.

The result of such geographic myopia was a neglect of creativity that was in some way vernacular, mundane, or amateur, but also a neglect of spaces beyond the urban – suburban towns, rural spaces, and ordinary places. As Edensor et al. (2009: 1) put it, "an understanding of vernacular and everyday landscapes of creativity honours the non-economic values and outcomes produced by alternative, marginal and quotidian creative practices". In response, closer considerations of the 'where' of creativity and the what, i.e. the beyond economic, have evolved and taken form in a recovery of the 'other' geographies of the creative economy as well as in the attention to vernacular creativities. If the former draw our attention to creative suburbias, the creative economy of small towns and rural places and creative clusters beyond the city, then the latter encourage us to look at the small scale or unexpected spaces and places such as garages, sheds, cafes, community centres, gardens, and homes as well as rubbish tips, allotments and online spaces (Edensor et al., 2009; Gregson et al., 2009; Gibson and Warren, 2016). The chapters in this volume situate making spatially in both local and historical contexts (see chapters by Thomas, Delyser and Edensor in particular), as well as in a variety of specialist locations such as makerspaces, studios and, indeed, sketchbooks (see chapters by Sjoholm, Rebecca Collins, DeSilvey and Ryan), as well as taking place in homes, backyards and hairdressers, community centres and churches (see chapters by Carr et al., Zoe Collins, Burke and Holmes). Moreover, these making practices are often situated in the context of contemporary environmental, social, development and labour concerns (see for example chapters by Carr et al., Burke, Straughan and Ocejo in particular).

This commitment to expanding where critical geographical engagement with creative practices takes place, also requires that we challenge those geographies of making that might settle out into accepted imagined geographies. Whether these concern differences between, for example, the creativity of urban and rural areas (the former associated with a creative economy, the latter with craft traditions as part of a good life), or between unseen amateur home-based creativities and more public professional forms of making.

To pick up, briefly, on just one of these strands – that of public/private binaries: knitting historically has taken place at home, however, recently the development of artistic and interventionist movements like yarnbombing signal its move into 'the public arena'. As Myzelev (2009: 161) observes, "similar to other activities such as embroidery, crocheting, and breastfeeding, it [knitting] allows women and in some instances men to bring their private hobbies to public spaces and thus reformulate, even if temporarily, the function of public areas such as cafes, buses, and libraries". Of course as Price (2015) and others note, not everyone who knits moves, or wants to move, their knitting into public spaces; there are complex spaces of domesticity and notions of femininity that are at work within and beyond gender, as well as considerations of class, race, and religion. Everyday life is increasingly inter-connected with digital spaces, particularly in the home, and as such demands that geographers pay attention to the geographies of making and the changing nature of domestic space. A top online site for home-based creative work is Etsy, known as 'e-bay for the hand-made'. By 2015 it had 15 million members and amassed $63 million in sales each month. Women make up over 90% of Etsy's sellers, many of whom have set up shops online to enable the negotiation of a family friendly work-life. A common narrative around 'the Etsy-seller' is the economically active woman who, on giving up her job to have children, sets up a micro-enterprise to combine contributing to household income and on-going participation in world of work with caring responsibilities. As Luckman (2013: 262) explains of Etsy "craft entrepreneurialism is more than an extension of thrifty housewifery and of 'making do' removed from the sphere of the monetary marketplace. Rather, it is precisely the kind of pro-am creative entrepreneurialism enabled by the social and economic expansion of the Internet." Feminist geographers, such as Carol Ekinsmyth (2011, 2014), have been concerned with gendered and emergent forms of entrepreneurial businesses "that are not merely (or even necessarily) located in the home, but creatively use the home, mother role and child oriented neighbourhood space(s) to do business" (Ekinsmyth, 2014: 1230). There is not the space to do justice here to the multiple, unsettled, spaces that the geographies of making might encompass. Yet, what is clear across this volume and wider disciplinary field, is that, firstly, sites of making are expanding beyond those specialist sites of the workshop, factory, studio and makerspace, and secondly, that practices of making often require that we keep open and challenge fixed imaginaries of places, in other words that ideas of home, of rural and urban, of public and private spaces are unsettled, remade through practices of making.

Maker bodies

We encounter a range of 'making' bodies across this volume. These are bodies that are learning to do; bodies that are tired and strained; bodies that are repulsed, that are skilled, that are satisfied 'in the flow', that are passionate, that are privileged; that are being coerced, frustrated and damaged (see especially Straughan, Ocejo, Collins, Holmes, this volume). Indeed, the body, that space closest in, is emerging as one of the most common sites in contemporary geo-graphical discussions of making practices. There are perhaps two key contexts that ground our understandings of these making bodies. On the one hand the wider turn to the body and practice. Within geography, at least, this has become coded through the advancing of non-representational geographies and those feminist geographies that have long placed bodily experience and knowledge at the fore (Anderson and Harrison, 2010). On the other, it is perhaps a re-visita-tion of labour geographies that have had a tendency to distance themselves from working bodies but which are now, to various degrees, attempting to recognise the viscerality and material practice of bodies who make, produce and craft (see for example discussions in Carr and Gibson, 2015, 2017). This speaks to, amongst other perspectives, the political work of geographers who 'follow the thing' in order to reveal the working conditions endured by those who produce our clothing and objects. Such a political commitment is often entwined with a desire to hand-make, or advocate the hand-made, because running throughout these geographies is a recognition of the labour, skill and intensity required of bodies to produce our stuff and how increasingly fast methods of production may damage these bodies (human and non-human; particularly the later when considering the environmental impacts of industry). In the case of both these fields of work we hope that engaging the geographies of making not only draws these two often disparate fields together but in doing so might offer new con-tributions to them.

Pioneered by phenomenologists and feminist geographers, and growing to encompass non-representational theory, geographical engagement with the body combines phenomenological and post-phenomenological engagements with being in the world with concerns around skill and habit, intersections of cognition and the affective and emotional, as well as with appreciations of the messy-fleshy nature of our bodies (Bissell, 2012; McCormack, 2014; Pickerill, 2015; Waitt and Stanes, 2015). The wider making literatures are no stranger to these approaches. Indeed, there is much geographers can learn by thinking through existing literature by sociologists, anthropologists and others that explores the phenomenological and embodied practice of making (O'Connor, 2007; Dudley, 2014; Ocejo, 2014b).

One of the primary dimensions of the bodies often present in discussions of making is that of the senses. So as Norris describes in her encounters with worn clothing:

> working on worn clothing involves engaging with their whole bodies, smelling its overwhelming odours released into the dusty warmth,

scanning its colours and patterns to assess its value, feeling the prickly wool and plastic acrylic, slippery linings and ridges of seams between practised hands.

(Norris, 2012: 41)

Making here encompasses a profound sensory intimacy, a "perceptual encounter with its invasive materiality". Such accounts of sensory engagement with materials in the making of things offer some of the richest contours of contemporary making discussions. Writing for example about craft breweries Thurnell-Read (2014: 8) suggests "while brewers did, to some extent, offer stock narratives of their entry into the trade, where their accounts became most energetic was when talking about the brewery as a space with an almost magical coming together of affective attachments, embodied processes and tangible sensory stimuli".

Importantly, the bodies that emerge through these literatures and across the chapters presented here are bodies that sense and are becoming skilled at sensing, through complex intersections of embodiment, materiality, matter, identity and skill, memory and habit. Furthermore, these are not the universal (white, masculine, able-bodied) bodies that populate much phenomenology, rather making bodies here emerge from their situations in intersectional social axes including gender, race, ethnicity and class, and are not only shaped by these intersectionalities, but also contribute to their on-going production.

Perhaps unsurprisingly, it is skill that has drawn much attention in the making literatures, and indeed many of the chapters collected here are careful to detail and qualify skill. For Sennett (2008: 10) "all skills, even the most abstract, begin as bodily practices". Whilst for Tim Ingold skill is a social and material practice, distributed across bodies and context; a form of knowledge that is not localised in individual bodies but rather assembled across social and geographical settings (Lea, 2009: 465). We witness this in the chapters across this text where cocktail makers, hair dressers, taxidermists and knitters reflect on their embodied skills, their learned habits, their sensing practices and their ability to improvise; all of which come to play a role in their making practices and command over the "mental, material, and physical aspects of the work" (Ocejo, 2010: 182). At times, discussions of making have fallen prey to a kind of fetishisation of the 'feeling of flow' (Csikszentmihalyi, 2008). This refers to the feeling of being so involved in an activity that time passes quickly and the line between task and world becomes blurred. This feeling requires a degree of skill, which allows the body to move with little conscious direction (Csikszentmihalyi, 2008). Here, whether through practices of learning to do, or a discussion of amateur creativities, or of the agency of unpredictable materials, such feelings of flow are not always achieved, but are at times actively resisted (see Delyser, Straughan, Holmes, Ocejo all this volume). We are keen therefore to highlight how the chapters compiled here also acknowledge complex, and potentially negative, bodily experiences of making, those that are weary, repetitive and arduous, and those for whom pursuit of a 'job well done' may have less than enchanting affective experiences.

To turn to our second framing for making bodies, that based on labour geography's working bodies, is to embrace a rather different corpus of scholarship. For rather than engaging with the body as the space closest in, labour geographies often discuss labouring bodies at a distance from their actual practices, experiences and techniques – as noted by McDowell (2015: 2). Labouring bodies are often assigned to particular categories of work (thinking about service work, or heavy industrial labour). In this vein, Carr et al., and Ocejo's chapters examine the contemporary recasting of creative and manufacturing work as craft labour. In doing so they explore the complex, potentially problematic, and varied landscape of work that is denoted and branded as 'craft' and the spaces such labour inhabits. Whilst Turnell-Read (2014: 48) has usefully deployed craft "as a means of accessing a set of wider insights into the nature of work and worker identity" we suggest doing so whilst engaging with how the neoliberalisation of such work has potentially exploitative outcomes, something ethnographers of creative labour more broadly have documented (Kucklich, 2005; Ocejo, 2014). In short, we suggest there are mutual benefits to bringing together these two very different fields of work on making bodies – labour geographies and those more social-cultural geographies of making. In other words, scholars of the geographies of making would do well to consider how the micropolitics of tasks, jobs and skill must be related to broader political arguments around labour, technological change, working bodies and identities.

Interlinked with this work on labouring bodies, and also of skill, are the discussions around professional and amateur making practices. Stephen Knott (2015: xvii) poses an important question when he asks "how we define amateur labour and how it is related to various philosophies of work". Indeed, he argues that "amateur craft is a space of production that departs from the normal conditions of everyday life, yet it is tangentially related to, and depends upon, its structures". In social terms, engaging with bodies that are making things brings to the fore questions of labour and work, and also leisure – for whom is making a hobby, done as an amateur, or as part of an everyday vernacular? For whom is making labour, work, a way to make a living? As this volume shows these lines are blurred, tense and often co-productive, and can reflect the effusive ways making penetrates everyday life, challenging binaries and geographical imaginaries.

This is reflected in more directed debates within craft and the arts. Crafts historian Glenn Adamson (2007: 140), for example, suggests, "in theory, hobbyists are beneath the notice of the expert. In practice, though, the line between the two is often a blurred one". The notion of 'amateur' then should not degrade the level of skill and expertise held by somebody engaged in creative practices that are in some way vernacular or everyday. As Helen Nicholson et al. (2016: 4) observe of the spaces of amateur dramatics and local theatre, it was found that "amateurism has become a by-word for poor-quality work, and a recurring theme in interviews is that amateurs resent the negative associations of the word". Indeed, "some people are offended by the word 'amateur', preferring 'local' or 'community' as a prefix" (Nicholson et al., 2015: 3). It therefore bears repeating that "the study of amateur craft practice must be more sensitive to its fluid and flexible status, and aware of how it is continually marginalised as a subject of study" (Knott, 2015: xv).

In short then, the bodies that populate geographies of making are complicated bodies, that both make and are made by individuals and collectives. They are messy, fleshy bodies, but are also thinking, feeling bodies that have habits, skills and memories, are happy or sad, invigorated or exhausted, inspired as well as strained and drained. They are bodies that are both produced by, but also productive of the particular social situations in which they are located. They are bodies which challenge and exceed these social situations as well as being cowed by them. These are in short making subjects, as well as, as we will see below, subjects shaped by making practices.

Matter and its ongoingness

Hair, flesh, spirits, paper, wood, stone and metal – these are just some of the matter and materials that figure throughout the chapters in this book (see in particular Holmes, Ocejo and Straughan, this volume). As Carr and Gibson (2015: 3) note, thinking about materials is central to making; "across the full spectrum of 'making cultures' are suggestions of sensibilities and dispositions that are centred on a deep and considered relationship with materials". As such, making scholarship resonates with wider humanities and social science research agendas that place concerns with matter and materiality to the fore. This interest reflects the sociological and political interest in addressing the power and agency of materials in the world. More particularly, these conversations are advancing work that recognises the vitality of matter and materialisms (Bennett, 2012).

As elsewhere, a number of the accounts in this book embroider the complex interactions of makers and materials. Akin to the discussions of butchery and taxidermy contained within this text, Chris Gibson (2014: 9) in his research on making leather cowboy boots explains the process as "grizzly contact with an assortment of nonhuman animals, feeling their dead skins, smelling them, and wearing them". He continues that animal traces are felt viscerally and are "responsible for the textual, sensual responses elicited both among bootmakers who manipulate skins into boots, and in tourists by touching, feeling and smelling hides" (Gibson, 2014: 9). Both Ocejo and Holmes (both this volume and Holmes, 2014), have written of the material challenges of working with hair, wherein the materiality of hair comes to stand for ethnicity and practices of self-care as well as for the intersection of the past and present interactions with hair-dressers, fashions and styles. In a rather different vein David Paton (2013) reflects the intensity of labour and working with stone by quoting a fellow quarry worker: "Fuck the granite! I am tired, I have a few more notches in my skin, a few more muscle fibres are anchored to bone, a few more tonnes of granite have passed though the filter of materiality and emerged forever unchanged". For Paton (2013: 1086): "the mundane reality of making, and thus of labour, is resolutely political, a geographical impera- tive, and a critical means of operating a meaningful relationship with this material life".

Throughout discussions of making, concerns with matter hover around its agentive nature, it acts back, it guides the maker, it is not simply material to be manipulated. Straughan, in her chapter in this volume, recounts her auto- ethnographic experiences of disgust, horror, as well as achievement when she encounters animal flesh as part of her taxidermy practice. Whilst, Patchett (2016: 414) describes how "the taxidermist is obliged to 'follow the material' and rhythmically respond to and negotiate its affordances while stitching back and forth across the cut". In short, making evolves as more than the simple human assertion of form onto static material. Rather, making is a co-production that sees a human maker interacting with, and being shaped by, the animate matter that is worked with (Ingold, 2013; Paton, 2013). As Ingold (2007: 7) articulates, "the forms of things are not imposed from without upon an inert substrate of matter, but are continually generated and dissolved within the fluxes of materials across the interface between substances and the medium that surrounds them". Such approaches to materials require that we foreground the on-going material making and shaping of the world.

To think of matter and materiality in this way is not only to appreciate the act of making that can produce a knitted sweater, a thrown pot, or a carved table, but is also to appreciate how an object might be remade – symbolically and more relevant here *materially* – multiple times, over its life course. As such, we are concerned to address practices beyond creation anew, orienting attention to practices of repair, maintenance and restoration (see Collins, DeLyser and DeSilvey et al. this volume). What must not go missing in geo- graphical discussions of making are such multiple makings, the overlapping lives of things that rework and extend biographies of objects via practices of mending, repairing, up-cycling or other ways of creatively re-working objects, including second-hand consumption practices, as well as those of restoration (Gregson and Crewe, 2003; Gregson et al., 2010; Bond, DeSilvey and Ryan, 2013; DeLyser and Greenstein, 2015, 2017).

Such practices of restoration and repair sit within a wider body of work that seeks to rethink discussions of rubbish and waste (Hawkins, 2010; Stras- ser, 1999). Indeed, as chapters by both Colllins and Carr et al. (this volume) explore, a disposition towards the world that seeks to repair rather than dis- pose of and buy anew is more likely to be cultivated if one is in possession of making skills. For Tim Dant (2010) the process of repair is not easy to plan or predict, it demands an emotional engagement that can adjust the human attention, sensitivity and effort to the objects being worked on. He continues that "the repertoire of gestures, the variable range of emotions and the flexible gathering of sensual knowledge needed for repair work are all distinctively human capacities" (Dant 2010: 18). The accounts offered by Delyser and Greenstein (including chapter 13 this volume, see also 2015, 2017) bear this out. Here restoration (whether concerned with neon signs or vintage cars) emerges as a thrilling and frustrating practice, and one that intersects a series of skills and techniques in knowing material properties, in knowing how to track the past stories of objects, in knowing people with skills and in being able and willing to offer the affective and emotional labours of love that are required. As DeLyser's and Greenstein's stories, but also those of Collins and De Silvey and Ryan in this volume make clear, tales of repair and restoration of objects are often firmly enmeshed within those of social relations, creating complex intersections of the on-going embodied skill, labour, and time needed to maintain craft, bodies and spaces.

What is being made?

If the geographies of making are multi-faceted, then the geographies being made *through* these practices are equally important. Many claims are made for the productive force of making practices, whether this be around economic regeneration, place-making or subject-forming. In chapters across the book we see making producing not only objects and experiences but also friendships (see Holmes and Burke); temporary places and events (see Edensor); human bodies and socio-material relations (see Straughan, Ocejo) and so on. In recent years, such concerns have shifted from telling wider scale stories of cities or places shaped through creative production, to a concern with the productive form of making's embodied and material dimensions. For some making becomes a geographical concern in the way local places, regions, institutions, and communities are shaped by histories and schooling of craftsmanship, skill and knowledge (Thomas et al., 2013; Hawkins, 2015, see also Thomas and Sjoholm this volume). For others, identity politics in creative making practices come to the fore, including reflecting on the role of practice and the agentive force of materials in shaping identity especially at specific points in the life course such as parent-hood or retirement (Bain, 2004, 2007; Yarwood and Shaw, 2010). What emerges is a sense in which making is a process of transformation – of materials, human competencies and spaces. As David Gauntlett (2011) and others (see Ravetz et al., 2013) have identified, this suggests making is a space of possibility and change, sustained engagement and collaboration.

Perhaps one of the most acknowledged debates of the power of making is its production of 'connections' and social relations. In a now oft repeated phrase David Gauntlett (2011) notes "making is connecting". This sentiment has increased the need for close empirical attention and critical reflection on the productive force of making, to how it produces the means and conditions through which alternative values and ways of living can be imagined and shared, and how practical examples for change can be defined and materialised. What follows traces three threads of geographies being made through this book; the making of self and social relations; the making of place and politics; and the making of environmental relations. What unites these threads is a concern here, but especially within the chapters, with querying *how* exactly *does* making *make* geographies? What are the scales in time and space of these geographies? How do they last and endure? And, what does it mean to talk about these geographies as being made?

Making self and social relations

Central to thinking about making practices beyond their economic value has been the recognition that practices of making fabricate our relationships with ourselves and with others, weaving the textures of our individual and collective identities. A classic example of this within the making literatures concerns the making of the modern home and the modern woman. As an extensive body of research, from oral histories to discourse analysis of magazines makes clear, there were strong intersections of making of the modern home with the making of feminine subjectivities, which included early twentieth-century gender debates and the promotion of well-being amongst both bored suburban housewives and the young emerging group of career women for whom society was concerned (Grace et al. 2009; Hackney, 2006). Home craft in the mid-twentieth century performed a feminine modernity binding modern and domestic in a wider ethic of handicrafts that interwove making and everyday life (Buckley, 1990; Hackney, 2006). As such 'making' was at one and the same time home-making, and was central to the production of the modern women and the ideal home. Whilst for some such control through craft was oppressive and stifling, for others craft was an important solace for the modern woman. An article in the *Modern Woman* observed, "what is needed is an outflow for nervous energy into other paths, and it probably is a self-protective instinct that makes a woman pick up her sewing or knitting while sitting still" (cited in Hackney, 2006).

The intersection of making and well-being should not be confined to the treatment of historical female neurosis. Knitting, like other lifestyle practices such as gardening and cooking, becomes newly significant as a means of creating or marking time as time for the self, outside of financial or familial responsibilities and duties (Hackney, 2006). Indeed, as David Gauntlett (2011: 106) has observed, to make is explicitly he says to feel "happy". At the heart of this happiness lies a sense of empowerment and connectivity, "making things shows us that we are powerful, creative agents – people who can really do things, things that other people can see, learn from, and enjoy. Making things is about transforming materials into something new, but it is also about transforming one's own sense of self" (Gauntlett, 2011: 435).

For others, the benefits of making lie in practices of doing, the therapeutic experiences offered by the role of tactility, texture, touch, and bodily intuition that acknowledges the role of materials and materiality in co-producing experience. If geographers have been interested in therapeutic landscapes

(Lea, 2009), they perhaps have been less concerned with therapeutic doings such as making practices. Importantly though, and as Zoe Collins explores in this volume, whilst practices such as knitting are leisure for some, for many women throughout the world such work continues to be underpaid, undervalued and unsatisfying in the context of their health, well-being and social-economic position (Clarke, 2016; Robertson, 2011; Groenveld, 2010; Dirix, 2014).

Making is often not only about a sense of self working alone, but also a sense of self that comes with knowing others. This is common to discussions of making, whether it be historical accounts of women's patchwork quilt parties or more contemporary discussions of online and off-line groups such as stich and bitch. More generally, geographers too have long recognised how certain groups take shape through the sharing of materialities, expertise and skills (Gregson and Crewe, 2003; Jupp, 2007; Askins and Pain, 2011; Hall and Jayne, 2015). Whilst Lorimer (2005: 86) noted the importance of exploring the habitual practices, intuitive acts and social protocols that draw together humans, objects and technologies.

If we are to actually understand making as connection however we need to attend to the kinds of connections that making cultivates, their form and their temporalities and spatialities. For David Gaunlett, at the heart of the intersections of making and happiness are positive associations. Happiness, he writes, is about

> family, community and well-being. It cannot be determined by a certain level of material comfort. Instead, it stems from having meaningful connections with others, and meaningful things to do. These (making) projects are especially valuable if they are not contained at the individual level but involve some form of sharing, co-operation or contribution to other people's well-being.
>
> (Gauntlett, 2011: 126)

This is the primary way in which making is connection, in which making has come to be associated with the creation of communities that form around shared practice, skill and material knowledge (Gauntlett, 2011). In short, 'making is connecting' because it is collaborative, participatory and feels good (in an embodied sense, but also expanded sense: more caring, more agentic in context of consumerist cultures) (see also Sennett, 2008). Crafts appear as a pleasurable social activity that embrace productive forms of leisure and that stand in opposition to the time-saving, alienating promises of consumer culture (Nathanson, 2013: 104). This is echoed by Hackney (2013: 187) who suggests "the great strength of amateur hobbyist practice is that it brings communities of interest together reflectively and reflexively through a shared love of 'making' and in the context of everyday life".

A lot of the contemporary discussion about making as connecting does so against the foil of internet, web 2.0 and consumer capitalism, with the latter as a source of social alienation whereas the former offer both possibilities but also risks in their reworking of social relations (Gauntlett, 2011; Burgess,

2006; Luckman, 2015). It is important however, that we do not overlook the historic intersections of making practices and social relations. Whether it was the communities of craft guilds, the female intergenerational relations built through shared crafting or the social relations created through quilting, there has historically been a strong sense that, especially for women, making is connecting. Yet also crucial is a need to expend more care than has perhaps currently been expressed in understanding the form and kind of these relations, this is not to privilege the deep and enduring connections, indeed transient interactions also serve important purposes, but is to point to the need to attend carefully to the nature of these encounters.

Friendship, as evolved in chapters here by Holmes especially, but to a lesser extent by Burke, can, it seems, be fabricated through these making practices. Paying attention to the embodied being together of friends through making illustrates the potential of making practices to create shared affective and emotional worlds to which friendship, is, as other scholars have explored, the 'social glue' (Bunnell et al., 2012). There are of course other ways that making intersects with friendship, most notably in practices of gifting carefully made objects. Turney (2012: 22) has explored the material and metaphorical relationships between knitting and parenthood. She argues, "knitting for one's children may be considered 'making love with needles'; it is a repetitive and painful exercise that is fruitless, thankless and endless; one that must be unpicked and reworked to achieve success". She draws links between knitting for family members and traditional feminine 'virtues' of caring and nurturing. For Turney (2012: 310), the assumption that knitting is necessarily more caring than other forms of production or consumption is problematic, not least because "the intention of the knitter, their thoughts, feelings, and emotions, demonstrated through the period spent knitting, may not be adequately communicated or received by the recipient".

There are then a series of ways in which making is connecting, through doing, gifting, talking and sharing, even if often these practices are not understood or communicated by all those involved. As Price (2015) discusses, however, it is important that we do not overlook how important solo making practices are and what this achieves for these individuals. Such a diversity of the values of making practices cautions against rushing to privilege connections forged through making together, but that an exploration of the form and kind of connections, their nature and durability, in short their spatialities and temporalities, is imperative.

Making place and politics

The relationship of making cultures to place identities is complex, located and often (re)imagined; whether it is tie-dye villages in rural China, traditions of lace making, or pottery in English towns and so on. Of late, however, such maker cultures have become folded within neoliberal policies that harness creative histories and manufacturing as part of formal place-making and branding agendas and policy (see Mould, 2015). Such that we see creative cities' mobilisation of creative economic practices as a means to stem industrial decline (e.g. US rust belt cities or cities in Northern England), or as a means to diversify away from a service or manufacturing economy (e.g. Shenzhen China). In rural places this might involve the cultivation of creative networks and hubs in communities that were once dependent on farming or tourism, or the exploitation of a particular heritage of making culture as a new form of place branding (Harvey et al., 2013). In such contexts, the role of creativity in place-making combines economic possibilities with social and material resources, whether the latter be new infrastructure – from buildings to gallery spaces – or whether it be aesthetically developed landscapes using plants and public art (Hawkins 2016; Mould 2015). The aim and accepted narrative in these instances is often to attract economically and socially wealthy individuals (the creative class) to areas that might not have been considered previously to be *rich* in cultural resources. Often these narratives have co-opted existing rich histories surrounding historical craft manufacturing, production and vernacular creativities and activities that have shaped places. This is not to dismiss the fact that whilst these heritages may become fodder for neoliberal policy makers, they are also central to the locations, histories and traditions themselves. Such economies are productive of the social and material fabric of places – having shaped its architectures, customs, infrastructure and broader cultural landscape. This argument is advanced by Edensor's contribution to this volume; in which it is made clear that organised place-making events need not only be part of neoliberal sets of practices, but rather may produce alternative community politics that mobilise the production of temporary places during festivals – a way of being otherwise. As he explores, place-making might foreground less the crafting of permanent additions to the material fabric of a place and might instead be about how the practices and products of making are productive of atmospheres.

Place-making represents only one dimension of the interaction of making with politics. Across the chapters within the book, politics and making are woven together in a range of ways. Whether that be the formal political structures of government and their funding regimes (expressed here through the provision of maker spaces and participatory arts organisations in chapters by Burke, Edensor and Collins), the more informal but still deeply felt politics of Guild organisations (see Thomas this volume), or the political elements of the NGO work discussed in Zoe Collins's chapter. Intersecting with these complex political contexts are, of course, the ways the practice of making is itself productive of the political.

Within the wider context of making, if not within the specific chapters of this book, there has been a clear alignment of making and politics through the rise of craftivism. Whilst forms such as yarnbombing or Guerrilla knitting have become increasingly popular (Price, 2015), textile craft more broadly has historically been used to creative, crafty and subversive ends (Parker, 1989). Notably, the Suffragettes' movement employed sewing skills to create protest banners; similarly, the AIDS memorial blanket deployed fabric to commemorate loss of life and to shift the terms of the political debate (Gambardella, 2011). However, at the time of writing, the last decade has seen an identifiable proliferation of creative practices under the banner of 'craftivism', not least in the world-wide production of millions of pink pussy hats that became in 2017 a symbol of women's rights and political resistance. For Greer (2008: 127) these practices work "by taking two seemingly disparate words that are negatively stereotyped in their own ways (craft can be seen as dull or old-fashioned, activism as violent or radical) and combining them to create a new word, 'craftivism' [that] strikes out into new territory". But it is far from easy and simple to accept these forms of craft practices. As Robertson (2011: 186) asks: "is it possible that the political effectiveness of radical craft practice relies inherently on the gendering of textile work? Is it possible, in other words, that knitting, embroidery, and quilting [being] used to make political change in some spheres requires their subjugation in others?" Of course one of the largest debates focuses around whether or not much of the female orientated textile work can be thought of in terms of feminist politics or not. Kelly (2014: 142) found examples of individual knitters and knitting communities that are clearly not feminist and are largely apolitical, and argues that scholars and others writing about knitting primarily as art and activism have been "overly optimistic" about the potential for knitting as a location for feminist politics (Kelly, 2014: 143). Such discussions should also be productively read alongside Collins's critiques in this volume of development programmes that too simplistically deploy making as part of development practices.

As such, whilst many claims can be made for the efficacy of making practices and politics, whether through contemporary or historical expressions of craftivism, as many of the chapters here explore, these claims often need careful grounding. Such grounding might occur, as it does in the chapters which follow, in specific examples to understand the form and kind of politics at work in these practices. These might not take the most obvious or activist forms, but may involve other forms of personal and wider sets of politics. It is to the possibilities of making for environmental politics that we now turn.

Making environmental encounters

In an era of large-scale global environmental change how we relate meaningfully to our environment has become an increasingly crucial question. It is not one that might, at least at first glance be obviously engaged with through making practices. Yet the handmade, in the UK context for example, is often loaded with nostalgic, morally charged connotations of certain kinds of historical good life. One that is lived close to the land, and where making was central to practices of self-sufficiency. Of course, such historical narratives were not without their challenges then and now. Particularly given that such historical relations and imaginations are often reinvigorated as part of the contemporary scene. We see craft and making, for example, associated with the slow movement more broadly, for example, slow food, slow cities, slow activism (see Pink, 2007; Hayes Conroy and Hayes Conroy, 2010). These practices offer 'alternative' ways of participating in consumption practices. In an age of global travel and instant communication the temporality of craft (often associated with a slower, mindful use of time) may be a precarious attempt to escape the "tyranny of the moment" and shape a new subject (Parkins, 2004: 436). As Jefferies puts it (2011: 237) "the pleasure of making things and the products of human hand, whether individually fashioned or collectively produced as part of social activities, enhance qualities of life and communication beyond the mundane, the superficial and the corporate". Indeed, in terms of fashion, hand-making can be viewed as "a return to a different relationship between fashion and consumption in which we see our clothes as long term investment pieces that speak of durability, love, attachment, quality and craft" (Crewe, 2013: 202). Yet, it is not just through an association with the 'good life' that we might explore the environmental encounters created through making practices.

A series of chapters in this volume explore how it is that making practices might reconstitute our environmental relations. Carr et al. (this volume, see also Carr and Gibson, 2015) observe the potential of making and mending prefaced by concerns with the precarity of our material world under the Anthropocene: "the ability to work with materials, and to make, repair or repurpose physical things, are vital skills, for a future where such resources become increasingly limited and extreme events related to a shifting climate are more common" (Carr and Gibson, 2015: 3). For Collins (this volume) the cultivation of maker habitus evolves a skill set across generations, cultivating repair and repurposing behaviours on a daily basis and as part of the practice of everyday life.

Other chapters in the book speak less to the process of making as an environmentally friendly behaviour and more to how practices of making may cultivate environmentally aware dispositions. Straughan, and to a lesser degree Ocejo, centralise how making practices require and develop attunement to non-humans. Straughan's account of learning taxidermy brings to light her corporeal intimacy and proximity with the bodies of non-humans. Something that happens in a very different context through Ocejo's discussion of butchery skills, and how they evolve a knowledge of animal life histories. These examples may cultivate knowledge and intimacy with non-humans, whether this is the same as cultivating environmentally aware behaviours, is unclear and, in Ocejo's case unlikely. What both examples do do, however, is illustrate how working in proximity to animal bodies (through making practices) may cultivate human-non-human intimacies. Along similar lines, Burke's chapter in this volume explores how the practices of knitting non-humans, in this case the very uncharismatic species of slugs and mice, might in fact attune people to their local environments. This is not, as she notes using participatory art frameworks, as simple (nor as linear) as suggesting that making is connecting with the environment. Rather, what is put forward is a critical account of the form and particular kinds of connections that are being built, the nature of the species engaged with, and how it is that making builds these relations.

In short then, making has all kinds of intersections with the shaping of environmental relations and behaviours. Some of which are more explicitly environmentally sensitive, sustainable practices, other facets of which are more philosophical. These involve less making behaviours that are directly sustainable actions, but rather the cultivation of dispositions and relations with the non-human that hold within them the potential to remake our environmental relations.

Making-unmaking-remaking

To return then to the ideas that opened this section; that the empirical and conceptual weave of discussion across this volume foregrounds the intersection of the geographies of making and the making of geographies. Such an inter-section requires a series of understandings that unmake and remake making, such that it evolves as a nimble and agile concept that refuses to settle out into accepted spaces and practices. Rather, the volume seeks to promote a sense of making that is expanded, and that requires us to be aware of how the specificities of different practices make and remake the very understandings of making itself. Making emerges as an active practice that concerns the un/re making of fixed and static spaces, identities and conceptual categories- public/ private, material/immaterial, representational/non-representational, professional/ amateur. What has not been sought here is a comprehensive review of the vast and diffuse field of making. Rather we have sought to reflect on how diverse literatures on making in the cross-disciplinary field, and across Geography's sub-disciplinary field, might enable a series of forms of unmaking in which the terms and understandings of makings' geographies are remade in productive tension.

Researching making

Research on making practices seems to demand that we rethink the kinds of epistemologies and ontologies that have more commonly marked out our research practices. Focusing research and critical geographical engagement on making has brought to the fore an urgency and methodological recognition that such interest requires challenging standard methods and research practices. What is needed are research and writing methods that make space for explorations of embodied and often unconscious practices, and that are able to take account of the messy, unpredictable and agentive materialities of making. The rise of making as a research topic has occurred concurrently with the turns to embodiment, affect, practice, and the acknowledgment of vibrant matter within geographical and wider scholarship. As such a rich range of methodological discussions has evolved which the explorations of methods of researching making must both take account of, but can also play into. In the discussion which follows we will examine how these methodological challenges evolve across the chapters in this volume and beyond.

The turn to embodied practices and concerns with vital and agentive matter has prompted, within and beyond geography, a series of methodological experiments. Interestingly, whilst Woodward (2016: 3) has written that so far "there has been very little methodological engagement with how qualitative methods might help us to understand materials and their properties, even as social scientists now argue for their centrality in the reproduction and breaking of social and material relations" the making ethnographies within this volume display an increasing attunement to research methods that enable us to *grasp* matter and materiality. Moreover, vocabularies are also evolving that enable us to present these vibrant matters in text, but also in spaces beyond the written: galleries, exhibitions, object production (see Hawkins, 2015). In the chapters that follow, the words of makers and their material sensitivities ensure that the affective force of maker–material relations are brought vividly to life on the page; be that the messy fleshy reality of taxidermy or butchery, the oily, frizzy, and course materiality of hair, or the unpredictable nature of plant based small batch spirits.

Whilst for many years geographers laboured under the misapprehension that people may find it difficult to talk about practices, this is not, as many chapters in this volume show, the case. In fact, many of the ethnographic approaches taken in this volume suggest that talking with makers (whether professional or amateur) reveals the rich languages and knowledge they have evolved for talking about their own practices, their skill sets, their relations to material, and their social circles and so on. Many of the studies here are characterised by a series of '*enriched*' interviews (Dowling et al., 2015) (for instance, talking about wool, with wool playing a part) allowing both for an exchange between researcher and maker expertise, and for the expertise of the material to be present (Woodward, 2016). In other cases conditions were created where making could be discussed in the doing. For example, Burke's chapter on knitting details how her own experience as a participatory artist and a knitter enabled her to create the conditions in which crafting could happen, and in so doing 'make' something as part of her research, in this case social and environmental relations, between knitters and between knitters and their local environments.

It is not only in talking to others that the practices, skills, and materialities of crafting practices become present, but also in the ways in which those studying making practices present auto-ethnographic representations of their own (see especially Straughan this volume). As Gibson and Carr (2017) note of their work on methods which 'animate making',

> our interest has been piqued by the many ways in which geographers working on making have brought their own life histories, bodies, materials and ideas to bear on fieldwork encounters, and in their writing. We are

particularly interested in how geographers have put their own bodies to work alongside those of participants, or have called upon longstanding making (and remaking) pursuits outside of academia, to extend understandings of the everyday, embodied accumulation of skill and tacit knowledge.

For many researchers interested in making what is required is a recognition of the entanglement of personal and professional identities. Often geographers, within and beyond this volume, have looked to their hobbies, skills, crafts or previous disciplinary training outside of geography for research topics and methods, for example: knitting, surfing and surfboard making, sculpture, taxidermy, painting (Mann, 2015; Warren and Gibson, 2014; Paton, 2013; Crouch, 2003). Yet there is a complex politics of knowledge, skill, and power when researchers study their own practices, or as is increasingly the case with creative and making practices, come to know a practice through learning to do (Hawkins, 2016).

Geographers have become assistant hairdressers (Holmes, 2014), butchers (Ocejo, 2014), sculptors (Paton, 2013), vine farmers (Krzywoszynska, 2015), and gardeners (Pitt, 2015). These apprenticeships are framed as learning by doing, becoming skilled, and gaining new expertise and material knowledges (O'Connor, 2007). As we have explored elsewhere however (Price 2015; Hawkins 2015), we need to be careful that we do not pretend that such processes are easy or straightforward revelations of experiences. Indeed, as Patchett writes of craftwork apprenticeships:

> when studying craftwork it requires not just the personal instruction of a good teacher and thus placing ourselves in the position of apprentice, it requires working with an ethic of the apprenticeship – an ethic that recognises that we need to be prepared to experiment and put ourselves and our theories at risk in order to produce methods that openly and creatively respond to our more than human, more than textual, multisensorial worlds.
>
> (Patchett, 2015: 92)

Such experimental and potentially 'risky' engagements of the researcher with a doing of creative practices is not new. Geographers have long employed creative and participatory methods to 'get at' certain knowledges that may only be appreciated by taking part. Sometimes, this means building upon existing skills, interests or disciplines to become, for example, an artist-geographer or geographer-poet (Crouch, 2003; Cresswell, 2012; Hawkins 2010, 2015). Whilst feminist geographers have long encouraged personal and embodied methodologies (see Longhurst, 1995), it seems creative practices are celebrated because there is something in doing and making that more traditional methods fall short on. To access these creative, sometimes haptic, knowledges the body becomes an increasingly important tool in research (Crang, 2003).

Importantly though, we should not overly festishise doing and being accomplished at doing (Hawkins 2015). Indeed many of the accounts of auto-ethnographic doings offered direct attention towards learning to do and the frustrations and failures (as well as the pleasures) inherent within this (see Straughan this volume). Richard Ocejo (2014a) discusses the substandard 'patties' he made as an intern in a New York butchers; the patties browned and became waste as he overworked the meat with unskilled hands. Similarly, Hawkins (2015) has discussed the evolution, values and failures of her poor drawing skills in her collaboration with an artist. In short, while we can note a series of experimental 'learnings to do' with respect to making practices, it is also vital that while valuing the knowledge produced through such practices, we do not fetishise professional or emerging amateur skill sets. Instead we need also to value what good ethnographic and interview work can do in terms of appreciating the multi-faceted geographies of making. As we explore in the conclusion, we are concerned that we continue to reflect on the challenges of researching making, suggesting this is one of several 'new territories' into which we would appreciate seeing making move.

Crafting new territories

As the accounts of the chapters detailed above suggest, this volume brings together diffuse senses of making. We start with a suite of chapters that foreground the geographies of making, beginning with Sjöholm's chapter on studios and sketchbooks, we then progress through Edensor's account of place-making and Thomas's discussion of Guild practices. As politics becomes a stronger theme within the chapters, we turn to Collins's consideration of making and development and Carr et al.'s observations on making for the Anthropocene. Picking up themes of work from their chapter, discussion unfolds in this direction through Ocejo's extended discussion of craft and labour practices, to take a closer look at the embodied, skilled making practices, themes picked on and developed further in the chapters by Holmes, Straughan and Burke which follow. Also running across these chapters are close attentions to the human, and human–non-human relations that are created through making, whether that be politics of friendship in Holmes and Burke's chapter or animal intimacies in Straughan's chapter. The final three chapters continue these themes of embodied, skilled making and interrogate making's connective potential through attentiveness to ideas of repair and restoration. With Collins exploring ideas of making and sustainability, DeSilvey and Ryan reflecting on ideas of mending, and DeLyser exploring concerns with restoration. This rich and diverse set of ideas offer much food for thought, and so in the conclusion of the volume we identify three strands that run across this volume that direct us towards key questions for future discussions. Firstly, concerns with the form of the transformational politics that making makes possible, secondly, issues around matter and material lives and finally, a proposal for care-full geographies of making. Importantly what unites all three, in common with the volume more generally, is a recognition that taking seriously the geographies of making can and must encompass and appreciate the promise and possibilities of the making of geographies.

References

Adamson, G. (2013) *The Invention of Craft*. London: Bloomsbury.

Adamson, G. (2007) *Thinking through Craft*. London: Berg.

Anderson, B. and Harrison, P. (2010) *Taking Place: Non-Representational Theory*. London: Routledge.

Anderson, C. (2013) *Makers: The New Industrial Revolution*. London: Random House.

Ardivisson, A., Malossi, G. and Naro, S. (2010) Passionate Work? Labour Conditions in the Milan Fashion Industry. *Journal for Cultural Research*, 4(3), 295–309.

Bain, A.L. (2007) Claiming Space: Fatherhood and Artistic Practice. *Gender, Place and Culture*, 14(3), 249–265.

Bain, A. (2004) Female Artistic Identity in Place: The Studio. *Social and Cultural Geography*, 5(2), 179–193.

Bissell, D. (2012) Habit Displaced: The Disruption of Skilfull Performance. *Geographical Research*, 51(2), 120–129.

Bond, S., DeSilvey, C. and Ryan, J. (2013) *Visible Mending: Everyday Repairs in the South West*. Axminster: Uniform Books.

Brace, C. and Johns-Putra, A. (2010) Recovering Inspiration in the Spaces of Creative Writing. *Transactions of the Institute of British Geographers*, 35(3), 399–413.

Bratich, Z. J. and Brush, M. H. (2011) Fabricating Activism: Craft-work, Popular Culture, Gender. *Utopian Studies*, 22, 234–260.

Brooks, M. (2010) The Flow of Action: Knitting, Making and Thinking. In Hemmings, J. (ed.) *In the Loop: Knitting Now*. London: Black Dog Publishing.

Buckley, C. (1990) *Potters and Paintresses: Women Designers in the Pottery Industries 1870–1955*. London: Women's Press.

Bunnell, T., Yea, S., Peake, L., Skelton, T. and Smith, M. (2012) Geographies of Friendships. *Progress in Human Geography*, 36(4), 490–507.

Burgess, J. (2006) Hearing Ordinary Voices: Cultural Studies, Vernacular Creativity and Digital Storytelling. *Continuum: Journal of Media and Cultural Studies*, 20(2), 201–214.

Buzek, M. (ed.) (2011) *Extra/Ordinary: Craft and Contemporary Art*. Durham, NC: Duke University Press.

Campbell, C. (2005) The Craft Consumer: Culture, Craft and Consumption in a Postmodern Society. *Journal of Consumer Culture*, 5(1), 23–42.

Carr, C. (2017) Maintenance and Repair beyond the Perimeter of the Plant: Linking Industrial Labour and the Home. *Transactions of the Institute of British Geographers*. https://doi.org/10.1111/trans.12183

Carr, C. and Gibson, C. (2017) Animating Geographies of Making: Embodied Slow Scholarship for Participant-researchers of Maker Cultures and Material Work. *Geography Compass*, 11(6).

Carr, C. and Gibson, C. (2015) Geographies of Making: Rethinking Materials and Skills for Volatile Futures. *Progress in Human Geography*, 40(3), 297–315.

Conradson, D. (2003) Geographies of Care: Spaces, Practices and Experiences. *Social and Cultural Geographies*, 4(4), 451–454.

Corkhill, B., Hemmings, J., Maddock, A. and Riley, J. (2014) Knitting and Well-being. *Textile: The Journal of Cloth and Culture*, 12(1), 34–57.

Crawford, M. (2009) *Shop Class as Soulcraft: An Inquiry into the Value of Work.* London: Penguin.

Crawford, M. (2010) *The Case for Working with your Hands: Or Why Office Work is Bad for us and Fixing Things Feels Good.* London: Penguin.

Creighton, M. R. (2001) Spinning Silk, Weaving Selves: Nostalgia, Gender, and Identity in Japanese Craft Vacations. *Japanese Studies*, 21(1), 5–29.

Cresswell, T. (2012) Value, Gleaning and the Archive at Maxwell Street, Chicago. *Transactions of the Institute of British Geographers*, 37(1), 164–176.

Crewe, L. (2013) Tailoring and Tweed: Mapping the Spaces of Slow Fashion. In Bruzzi, S. and Church Gibson, P. (eds) *Fashion Cultures: Theories, Explorations, Analyses.* London: Routledge.

Crouch, D. (2003) Spacing, Performing, and Becoming: Tangles in the Mundane. *Environment and Planning A*, 35(11), 1945–1960.

Crouch, D. and Matless, D. (1996) Refiguring Geography: Parish Maps of Common Ground. *Transactions of the Institute of British Geographers*, 21(1), 236–255.

Csikszentmihalyi, M. (2008) *Flow: The Psychology of Optimal Experience.* New York: Harper Perennial.

Dant, T. (2010) The Work of Repair: Gesture, Emotion and Sensual Knowledge. *Sociologicial Research Online*, 15(3).

DeLyser, D. (2001) 'Do you really live here?' Thoughts on Insider Research. *The Geographical Review*, 91, 441–453.

DeLyser, D. (2014) Tracing Absence: Enduring Methods, Empirical Research and a Quest for the First Neon Sign in the USA. *Area*, 46(1), 40–49.

DeLyser, D. (2015) Collecting, Kitsch and the Intimate Geographies of Social Memory: A Story of Archival Autoethnography. *Transactions of the Institute of British Geographers*, 40(2), 209–222.

DeLyser, D. and Greenstein, P. (2015) "Follow That Car!" Mobilities of Enthusiasm in a Rare Car's Restoration. *The Professional Geographer*, 67(2), 255–268.

DeLyser, D. and Greenstein, P. (2017) The Devotions of Restoration: Enthusiasm, Materiality, and Making Three "IndianMotocycles" New. *Annals of the Association of American Geographers.* http://dx.doi.org/10.1080/24694452.2017.1310020

Dirix, E. (2014) Stitched Up – Representations of Contemporary Vintage Style and the Dark Side of the Popular Knitting Revival. *Textile: The Journal of Cloth and Culture*, 12(1), 86–99.

Dowling, R., Lloyd, K. and Suchet-Pearson, S. (2015) Qualitative Methods 1: Enriching the Interview. *Progress in Human Geography*, 40(5), 679–686.

Dudley, K. (2014) *Guitar Makers: The Endurance of Artisanal Values in North America.* Chicago: University of Chicago Press.

Edensor, T. (2011) Entangled Agencies, Material Networks and Repair in a Building Assemblage: The Mutable Stone of St Ann's Church, Manchester. *Transactions of the Institute of British Geographers*, 36(2), 238–252.

Edensor, T., Leslie, D., Millington, S. and Rantisi, N. (eds) (2009) *Spaces of Vernacular Creativity: Rethinking the Cultural Economy.* London: Routledge.

Ekinsmyth, C. (2014) Mothers' Business, Work/Life and the Politics of 'Mumpreneurship'. *Gender, Place & Culture*, 21(10), 1230–1248.

Ekinsmyth, C. (2011) Challenging the Boundaries of Entrepreneurship: The Spatialities and Practices of UK 'Mumpreneurs'. *Geoforum*, 42(1), 104–114.

Faiers, J. (2014) Knitting and Catastrophe. *Journal of Cloth and Culture*, 12(1), 100–109.

Fallows, J. (2016) Why the Maker Movement Matters: The Tools, the Revolution. *The Atlantic*, 5th June 2016. https://www.theatlantic.com/business/archive/2016/06/why-the-maker-movement-matters-part-1-the-tools-revolution/485720/ [last accessed 30/9/2017]

Fisher, C. (1997) 'I bought my first saw with my maternity benefit'. Craft Production in West Wales and the |Home as the Space of Re (Production). In Cloke, P. and Little, J. (eds) *Contested Countryside Cultures: Rurality and Socio-cultural Marginalization*. London: Routledge.

Fisher, T. and Botticello, J. (2016) Machine-made Lace. The Spaces of Skilled Practices and the Paradoxes of Contemporary Craft Production. *Cultural Geographies*. http://journals.sagepub.com/doi/abs/10.1177/1474474016680106.

Gambardella, S. J. (2011) Absent Bodies: The AIDS Memorial Quilt as Social Melancholia. *Journal of American Studies*, 45(2): 213–226.

Gauntlett, D. (2011) *Making is Connecting: The Social Meaning of Creativity from DIY and Knitting to YouTube and Web 2.0*. London: Polity Press.

Gelber, S. M. (1997) Do-it-yourself: Constructing, Repairing and Maintaining Domestic Masculinity. *American Quarterly*, 49: 66–112.

Gibson, C. (2016) Material Inheritances: How Place, Materiality and Labor Process Underpin the Path-dependent Evolution of Contemporary Craft Production. *Economic Geography*, 92, 61–86.

Gibson, C. and Warren, A. (2016) Resource-sensitive global production networks: Reconfigured geographies of timber and acoustic guitar manufacturing. *Economic Geography*, 92, 430–454.

Gibson, C., Carr, C. and Warren, A. (2012) A Country that Makes Things? *Australian Geographer*, 43(2), 109–113.

Grace, M., Gandolfo, E. and Candy, C. (2009) Crafting Quality of Life: Creativity and Well Being. *Journal of the Association for Research on Mothering*, 11(1): 239–250.

Graham, S. and Thrift, N. (2007) Out of Order – Understanding Repair and Maintenance. *Theory, Culture and Society*, 24(1), 1–25.

Greer, B. (2008) *Knitting for Good*. London: Trumpeter Books.

Gregson, N. and Crewe, L. (2003) *Second-hand Cultures*. London: Berg Publishers.

Gregson, N., Crang, M., Ahamed, F., Akhter, N. and Ferdous, R. (2010) Following Things of Rubbish Value: End-of-life Ships, 'Chock-chocky' Furniture and the Bangladeshi Middle Class Consumer. *Geoforum*, 41(6), 846–854.

Gregson, N., Metcalfe, A. and Crewe, L. (2009) Practices of Object Maintenance and Repair: How Consumers Attend to Consumer Objects within the Home. *Journal of Consumer Culture*, 9, 248–272.

Groeneveld, E. (2010) Join the Knitting Revolution: Third-wave Feminist Magazines and the Politics of Domesticity. *Canadian Review of American Studies*, 40(2), 259–277.

Hackney, F. (2006) Using your Hands for Happiness: Home Craft and Make Do and Mend in British Women's Magazines in the 1920s and 1930s. *Journal of Design History*, 19(1), 23–38.

Hackney, F. (2013) Quiet Activism and the New Amateur: The Power of Home and Hobby Crafts. *Design and Culture*, 5(2), 169–193.

Hallam, E. and Ingold, T. (2008) *Creativity and Cultural Improvisation*. London, Oxford: Berg.

Hallam, E. and Ingold, T. (2014) *Making and Growing*. Farnham: Ashgate.

Hall, S. M. and Jayne, M. (2015) Make, Mend and Befriend: Geographies of Austerity, Crafting and Friendship in Contemporary Cultures of Dressmaking in the UK. *Gender, Place & Culture*. [ahead-of-print], 1–19.

Harvey, D., Hawkins, H. and Thomas, N. (2013) Creative Clusters beyond the City. *Geoforum*, 43(3), 529–539.

Hawkins, H. (2010) Turn your Trash into... Rubbish, Art and Politics. Richard Wentworth's Geographical Imagination . *Social and Cultural Geography*, 11(8), 805–827.

Hawkins, H. (2015) *Creative Geographies: Geography, Visual Art and the Making of Worlds*. London: Routledge.

Hawkins, H. (2016) *Creativity: Live, Work, Create*. London: Routledge.

Hayes Conroy, A. and Hayes Conroy, J. (2010) Visceral Geographies: Mattering, Relating, and Defying. *Geography Compass*, 4(9), 1273–1278.

Heatherwick, T. and Rowe, M. (2015) *Thomas Heatherwick: Making*. London: Thames and Hudson.

Hemmings, J. (2010) *In the Loop: Knitting Now*. London: Black Dog Publishing.

Hitch, M. (2013) *The Maker Movement Manifesto: Rules for the Innovation in the New World of Crafters, Hackers, and Tinkerers*. New York: McGraw-Hill.

Holmes, H. (2014) Chameleon Hair: How Hair's Materiality Affects its Fashionability. *Critical Studies in Fashion & Beauty*, 5(1), 95–110.

Holmes, H. (2015) Transient Craft: Reclaiming the Contemporary Craft Worker. *Work, Employment & Society*, 29(3), 479–495.

Hracs, B. and Leslie, D. (2013) Aesthetic Labour in the Creative Industries: The Case of Independent Musicians in Toronto, Canada. *Area*, 46, 66–73.

Ingold, T. (2007) Materials against Materiality. *Archaeological dialogues*, 14(1), 1–16.

Ingold, T. (2009) The Textility of Making. *Cambridge Journal of Economics*, 34, 91–102.

Ingold, T. (2013) *Making: Anthropology, Archaeology, Art and Architecture*. London: Routledge.

Ingold, T. (2014) That's enough about ethnography! *HAU: Journal of Ethnographic Theory*, 4, 383–395.

Jefferies, J. (2011) Loving Attention: An Outburst of Craft in Contemporay Art. In Buzek, M. (ed.) *Extra/Ordinary: Craft and Contemporary Art*. Durham, NC: Duke University Press, 222–241.

Kelly, M. (2013) Knitting as a feminist project? *Women's Studies International Forum*, 44, 133–144.

Knott, S. (2015) *Amateur Craft: History and Theory*. London: Bloomsbury.

Korn, P. (2017) *Why we Make Things and Why it Matters: The Education of Craftsmen*. London: Vintage.

Krzywoszynska, A. (2015) What farmers know: experiential knowledge and care in vine growing. *Sociologia Ruralis*, 52(2), 289–310.

Kucklich, J. (2005) Precarious Playbour: Modders and the Digital Game Industry. *Fibreculture*, 5(11).

Lang, D. (2013) *Zero to Maker: Learn (Just Enough) to Make (Just About) Anything*. London: Maker Media.

Lea, J. (2009) Becoming Skilled: The Cultural and Corporeal Geographies of Teaching and Learning Thai Yoga Massage. *Geoforum*, 40(3), 465–474.

Leadbeater, C. and Miller, P. (2004) The Pro-am Revolution: How Enthusiasts are Changing our Economy and Society. *Demos* [Online]. Accessed from: http://www. demos.co.uk/files/proamrevolutionfinal.pdf [Accessed 25/7/2015]

Lindtner, S. (2015) Hacking with Chinese Characteristics: The Promise of the Maker Movement Against Chinese Manufacturing Culture. *Science, Technology and Human Values*, 40(5), 854–879.

Longhurst, R. (1995) Viewpoint: The Body and Geography. *Gender, Place and Culture*, 2(1), 97–106.

Luckman, S. (2013) The Aura of the Analogue in a Digital Age: Women's Crafts, Creative Markets and Home-based Labour after Etsy. *Cultural Studies Review*, 19, 249–270.

Luckman, S. (2015) *Craft and the Creative Economy.* Basingstoke: Palgrave Macmillan.

Mackenzie, A.F.D. (2006c) 'Leinn Fhàin Am Fearann?' (the Land Is Ours): Re-Claiming Land, Re-Creating Community, North Harris, Outer Hebrides, Scotland. *Environment and Planning D: Society and Space*, 24(4), 577–598.

Mann, J. (2015) Towards a Politics of Whimsy: Yarn Bombing the City. *Area*, 47, 65–72.

McCormack, D. P. (2008) Geographies for Moving Bodies: Thinking, Dancing, Spaces. *Geography Compass*, 2(6), 1822–1836.

McCormack, D. P. (2014) *Refrains for Moving Bodies.* Durham, NC: Duke University Press.

McCabe, M. and de Waal Malefyt, T. (2013) Creativity and Cooking: Motherhood, Agency and Social Change in Everyday Life. *Journal of Consumer Culture*, 15(1), 48–65.

McMorrin, C. (2012) Practising Workplace Geographies: Embodied labour as method in human geography. *Area*, 44, 489–495.

Minahan, S. and Cox, J. W. (2011) The Inner Nana, the List Mum and Me: Knitting Identity. *Material Culture Review/Revue de la culture matérielle*, 72. https://journals. lib.unb.ca/index.php/MCR/article/view/18719/20463

Moore, M. and Prain, L. (2009) *Yarn Bombing: The Art of Crochet and Knit Graffiti.* Vancouver: Arsenal Pulp Press.

Morozov, E. (2014) Making It: Pick up a Spot Welder and Join the Revolution. *New Yorker*, 13th June 2014, http://www.newyorker.com/magazine/2014/01/13/ma king-it-2

Mould, O. (2015) *Urban Subversion and the Creative City.* Routledge: London.

Mountz, A., Bonds, A., Mansfield, B., Lloyd, J., Hyndman, J., Walton-Roberts, M., … Curran, W. (2015) For Slow Scholarship: A Feminist Politics of Resistance through Collective Action in the Neoliberal University. *ACME International E-Journal for Critical Geographies*, 14, 1235–1259.

Myzelev, A. (2009) Whip your Hobby into Shape: Knitting, Feminism and Construction of Gender. *Textiles: A Journal of Cloth and Culture*, 7(2), 148–163.

Nathanson, E. (2013) *Television and Postfeminist Housekeeping: No Time for Mother.* London: Routledge.

Nicholson, H. (2015) Absent Amateurs. *Research in Drama Education*, 20(3), 263–266.

Norris, Lucy. (2012) Shoddy Rags and Relief Blankets: Perceptions of Textile Recycling in North India. In Alexander, C. and Reno, J. (eds) *Economies of Recycling: The Global Transformation of Materials, Values and Social Relations.* London: Zed Books, 35–58.

O'Connor, E. (2005) Embodied Knowledge: Meaning and the Struggle Towards Proficiency in Glassblowing. *Ethnography*, 6(2), 183–204.

O'Connor, E. (2006) Glassblowing Tools: Extending the Body towards Practical Knowledge and Informing a Social World. *Qualitative Sociology*, 29(2) 177–193.

O'Connor, E. (2007) Embodied Knowledge in Glassblowing: The Experience of Meaning and the Struggle towards Proficiency. *The Sociological Review, 55*, 126–141.

Ocejo, R. E. (2010) What'll it be? Cocktail Bartenders and the Redefinition of Service in the Creative Economy. *City, Culture and Society*, 1(4), 179–184.

Ocejo, R. E. (2012) At your Service: The Meanings and Practices of Contemporary Bartenders. *European Journal of Cultural Studies*, 15(5), 642–658.

Ocejo, R. E. (2014a) Show the Animal: Constructing and Communicating New Elite Food Tastes at Upscale Butcher Shops. *Poetics*, 47, 106–121.

Ocejo, R. E. (2014b) *Upscaling Downtown: From Bowery Saloons to Cocktail Bars in New York City*. Princeton: Princeton University Press.

Orton-Johnson, K. (2014) Knit, Purl and Upload: New Technologies, Digital Mediations and the Experience of Leisure. *Leisure Studies*, 33(3), 305–321.

Parker, R. (1989) *The Subversive Stitch: Embroidery and Making of the Feminine*. London: Women's Press.

Patchett, M. (2015) Witnessing Craft: Employing Video Ethnography to Attend to the More-than-human Craft Practices of Taxidermy. *Video Methods: Social Science Research in Motion*. London: Routledge, pp. 71–94.

Patchett, M. (2016) The Taxidermist's Apprentice: Stitching Together the Past and Present of a Craft Practice. *Cultural Geographies*, 23, 401–419.

Patchett, M. (2017) Taxidermy Workshops: Differently Figuring the Working of Bodies and Bodies at Work in the Past. *Transactions of the Institute of British Geographers*, 42(3), 490–504.

Patchett, M. and Foster, K. (2008) Repair Work: Surfacing the Geographies of Dead Animals. *Museum and Society*, 6(2), 98–122.

Paton, D. (2013) The Quarry as Sculpture. *Environment and Planning A*, 45(5), 1070–1086.

Pentney, B. A. (2008) Feminism, Activism, Knitting: Are the Fibre Arts a Viable Mode for Feminist Political Action. *Thirdscape: A Journal of Feminist Theory and Culture*, 8(1). [Accessed Online].

PickerillJ. (2015) Bodies, Building and Bricks: Women Architects and Builders in Eight Eco-communities in Argentina, Britain, Spain, Thailand and USA. *Gender, Place and Culture*, 22(7), 901–919.

Pink, S. (2007) Sensing Cittàslow: Slow Living and the Constitution of the Sensory City. *The Senses and Society*, 2(1), 59–77.

Pink, S., Mackley, K. L. and Moroşanu, R. (2015) Hanging out at Home: Laundry as a Thread and Texture of Everyday Life. *International Journal of Cultural Studies*, 18(2), 209–224.

Podkalicka, A. and Potts, J. (2013) Towards a General Theory of Thrift. *International Journal of Cultural Studies*, 17(3), 227–241.

Pollock, G. (1999) *Differencing the Canon: Feminist Desire and the Writing of Art's Histories*. London and New York: Routledge.

Price, L. (2015) Knitting and the City. *Geography Compass*, 9(2), 81–95.

Ravetz, A., Kettle, A. and Felcey, H. (2013) *Collaboration Through Craft*. London: Bloomsbury.

Riley, J. (2013) The Benefits of Knitting for Personal and Social well-being in Adulthood: Findings from an International Survey. *The British Journal of Occupational Therapy*, 76(2), 50–57.

Robertson, K. (2011) Rebellious Doilies and Subversive Stitches: Writing a Craftivist History. In M.E. Buszek (ed.), *Extra/Ordinary: Craft and Contemporary Art*. Durham and London: Duke University Press, 184–203.

Sennett, R. (2008) *The Craftsman*. London: Penguin.

Sjöholm, J. (2012) *Geographies of the Artist's Studio*. London: Squid and Tabernacle.

Sjöholm, J. (2014) The Art Studio as Archive: Tracing the Geography of Artistic Potentiality, Progress and Production. *Cultural Geographies*, 21(3), 505–514.

Stalp, M. C. (2007) *Quilting: The Fabric of Everyday Life*. London: Berg.

Straughan, E. (2015) Crafts of Taxidermy: Ethics of Skin. *GeoHumanities*, 1(2), 363–377.

Thomas, N. J., Harvey, D. C. and Hawkins, H. (2013) Crafting the Region: Creative Industries and Practices of Regional Space. *Regional Studies*, 47(1), 75–88.

Thurnell-Read, T. (2014) Craft, Tangibility and Affect at Work in the Microbrewery. *Emotion, Space and Society*, 13, 46–54.

Tolia-Kelly, D. P. (2013) The Geographies of Cultural Geography. III Material Geographies, Vibrant Matters and Risking Surface Geographies. *Progress in Human Geography*, 37(1), 153–160.

Turney, J. (2009) *The Culture of Knitting*. Oxford: Berg.

Turney, J. (2012) Making Love with Needles: Knitted Objects as Signs of Love? *Textile: Journal of Cloth and Culture*, 10(3), 302–311.

Waitt, G. and Stanes, E. (2015) Sweating Bodies: Men, Masculinities, Affect, Emotion. *Geoforum*, 59, 30–38.

Warren, A. and Gibson, C. (2014) *Surfing Places, Surfboard Makers: Craft, Creativity and Cultural Heritage in Hawai'i*. Hawai'i: University of Hawai'i Press.

Whatmore, S. (2006) Materialist Returns: Practising Cultural Geography in and for a More-than-human World. *Cultural Geographies*, 13(4), 600–609.

Williams, R. (2014) *Keywords*, London: Penguin.

Woodward, S. (2016) Object Interviews, Material Imaginings and 'Unsettling' Methods: Interdisciplinary Approaches to Understanding Materials and Material Culture. *Qualitative Research*, 16(4), 359–374.

Yarwood, R. and Shaw, J. (2010) 'N'gauging' Geographies: Craft Consumption, Indoor Leisure and Model Railways. *Area*, 42(4), 425–433.

2 Making bodies, making space and making memory in artistic practice

Jenny Sjöholm

Introduction

Artistic processes and creative making have, in recent years, attracted increasing attention in social and cultural geography with researchers focusing on the variety of ways in which artists experience and explore the environments which surround them, and how these spaces relate to their creative making (e.g. Bain, 2003; Crouch and Toogood, 1999; Hawkins, 2010; Morris and Cant, 2006; Rogers, 2011). This research has raised a series of important questions about the contexts and spatialities artistic creative processes rely upon. Against this background I suggest, there remain important questions about the role of the traditional artistic production space of the art studio and the making that unfolds within. In (re)visiting the studio I argue that we should follow Hawkins' (2010) suggestion that art making and creativity be analysed as a skilled, material and embodied practice.

In this chapter the contemporary art studio is viewed as a vital space of artistic making. The studio is approached as an important knowledge space, an archive and material repository, where artists store but also explore their personal collections of resources, memories and materialities through their creative processes. Inspired by Latour and Woolgar's (1979) laboratory studies, I approach artists' making through focusing on artists' knowledge practices and in particular the use of sketchbooks as a fundamental knowledge device in their creative processes. Using a practice oriented empirical examination, the chapter focuses on how artists come to perceive but also construct their own workspaces – here (most often) the art studio and the ecology of studio materials – and how they actively use these spaces and materialities.

For artists today, as always, the contexts within which their embodied practices take place are important for us to understand. There is a concern in geography that focusing on embodied practice means we are left with abstract, decontextualized and ahistorical accounts. I would make the case here that exploring making as an embodied practice means we engage with and animate matters in the context of wider histories and places. By understanding making primarily as an embodied practice we take seriously the life worlds of artists, we avoid abstract aggregations, we get closer to the often historically situated events that orient these practices, situated place-based individuals, networks and histories.

Although artists working in art studios has been a long convention within the making of art, there has been renewed interest in the role of 'the studio' as a space of creation and artistic production in scholarly work (Farías and Wilkie, 2016; Jacob and Grabner, 2010; Davidts and Paice, 2009) in illustrated books (Amirsadeghi, 2012; Madestrand, 2012) as well as being a popular focus for art exhibitions and exhibition texts (Waterfield, 2009; Major, 2012). Accounts in geography have so far focused on the relationship between the art studio and the construction of artistic identity and identity politics (Bain, 2003, 2005), on the studio as a marketized space (Sjöholm and Pasquinelli, 2014), and on the art studio as a geographical venue for knowledge and imagination (Daniels, 2011). Building on the ideas presented above the rest of the chapter attempts to advance the idea, put forward by Daniels (2011), of the studio as an enhanced micro-space of artistic making, knowledge and learning. The studio is here viewed as essentially an imagination and knowledge chamber where artists engage in practice based on material resources, knowledge, learned scholarship; a civilized pursuit based on learning (Daniels, 2011; Chare, 2006; Buren, 1970/71).

I approach the studio as a space of collection, categorization and documentation. The studio is where objects, devices and documents are placed as a way to mark an end to work processes; as well as it is a space where art originates or is reinvented. It is a space where things are stored, made, handled and begun. However, in its particular set-up there is a creative limitation: there is an agentive force and a limiting order of materials that authorize and command the future development of artistic processes. Making and thinking in the modern art studio is an archival as well as practice based process. Artists develop their own individual ways of practicing art in contexts that are structured by the presence of their own plans and collected materials. The generated knowledge and skills of the studios are properties of the whole human-organism-person that have emerged through the history of his or her involvement with that environment. The studio materialities, the collected materials, sketches, memories and references are fundamental resources in the studied artists' making of art.

This chapter is based on in-depth studies on artists' practice, methodologies and embodied knowledge: work that starts with a focus on the situated individual artist, on the artist as a worker and creator. Drawing on forty interviews and ethnographic fieldwork conducted with London-based visual artists inside and outside the workspace of the art studio, this work is of explorative rather than exhaustive character and the discussion is based on selected examples of artistic making processes in the studio. The chapter draws on cultural geographies and socio-material perspectives of making and learning to argue that micro-spatial and material perspectives are important to understanding artists' working and making practices and 'making geographies'.

Contemporary studio based making

The studio has long been a central feature of how we perceive artists' working lives but the contemporary studio can, of course, take a variety of forms and have many functions depending on the intention of the artist and the resources at their disposal (Wainwright, 2010). However, in recent times the studio's role has been in question and it has been suggested may, to some extent, even be in the process of being abandoned for other production modes and sites (e.g. Hawkins, 2012; Alpers, 2010; Buren, 1970/71). As Hawkins points out, 'an enduring feature of art's 20th-century expansion is a shift in the sites of art's production and consumption' (2012: p.56). And as stressed by Alpers: 'Painting made by someone alone in the studio is out... [whilst] art produced for or in public spaces – from earthworks and public sculpture or assemblages to photography, videos, graffiti, and work produced on gallery walls – is in' (Alpers, 2010: p.126).

Furthermore, the studio is also still somewhat a 'room of privilege' (Grabner, 2010) and as Bain's study of contemporary women visual artists from Toronto shows, the studio as an artistic workspace is a 'hard-won resource' (Bain, 2003). This chapter is focused on individual professional artists that are continuously facing change and challenge related to their professional life and career aspirations. Like the majority of cultural workers, artists are situated in fundamental uncertainty and institutional ambiguity due to volatile labour markets and a transition to precarious work or post-Fordist regimes of production as well as increased competition (Seijdel, 2009; Gill and Pratt, 2008; Ross, 2008).

Despite indications of the shifting nature of artistic sites, spaces and modes of production, this chapter rests on the premise that the art studio remains a fundamental space of contemporary artistic production, but that what might be understood as a rather traditional model of studio based artistic production, with accompanying notions of artistic creation as the expression of individual talent (Hughes, 1990), needs to be re-visited and studied through critical engagement. This chapter analyses the relationship between the situated skilled maker and the material resources and practices involved in the making of art and the art studio, through a focus on embodied processes of formation and making. Whilst spaces of the art world such as spaces of display and artist collectives have been studied (Markusen, 2006; Mackenzie, 2006; Rogers, 2011), little critical interest has been directed towards the micro-spatialities of artists' making and the artists' studios in terms of close accounts of situated individuals' actual making practices and studio work processes (Graw, 2003). Equally relatively little attention has been paid to how artists' experiences and work are affected or shaped by the resources, knowledge, devices, skills and materiality that take shape in these spaces.

The artists in this study are professional artists that have figured out a way to have a career in art but this does not mean that all of the artists have the opportunity to have their own permanent studio. Because of their uncertain professional situation, artists are often forced to adapt to alternative work-spaces and practices that sometimes challenge the importance of the studio. As Alpers writes, 'the challenge to the studio in recent times is not Warhol's factory model but the greater world outside it' (2010: p.127). For those artists in this study without their own permanent studio they often created alter-native 'studios': by getting access to studios through temporary residencies or through reimaging their kitchens or garages as making spaces. However, for all of them having their own studio is what they aim and plan for. Having access to a studio is arguably a way for contemporary visual artists to main-tain their individual making and learning, as well as it is arguably central to how artists stabilize and participate in a professional discourse and identity (Bain, 2004, 2005; Buren, 2007; Jacob, 2010).

To have your own studio space seems to be a strategy for artists to carve out a space in which they can seek and maintain continuity and stability in managing their work, creativity and professional development. Without idea-lizing the role of the studio, it is perhaps safe to say that the studio is more important for contemporary artists than ever before. Whilst the nature of the studio keeps changing and the ability of artists to afford one is constantly challenged, there is also still a prevalent studio based model of production present in artistic making and even artists that do not have a studio are making art with the studio and a studio based production model in mind. The studio is still an important space to understand as it is a condition and an instrument for creativity, as well as being a state of mind. As Jacob puts it, the studio 'remains a place where artists go or long to return to, something they have or seek to obtain' (2010: p. xii).

The sketchbook – a studio based knowledge device

The studio is not a closed off space but a making space that is interconnected and dependent on other sites of artistic practice and production. Artistic making is partly a gathering process – it is about soaking up experiences, impressions, objects, references and narratives from journeys outside the workspace of the studio (e.g. Sachs Olsen and Hawkins, 2016). To the artists of this study, their sketchbook was a device that linked studio based practice and such itinerant practice. Sketching and drawing, it is suggested here, can be approached as a portable medium, a fundamental exploratory tool or device through which artists not only explore the world inside as well as outside the studio, but also develop new ideas and prepare new work.

For these artists, whether it was spread sketches, collected sketches in a sketchbook, notes in a notebook or digital records, this material often had the same destiny: they will eventually be brought back to the art studio, or other similar work space, for storage, further work and making. The sketch-books consist of reflections and interpretations of their past experiences, but they are also manuscripts and instructions for future investigation. The sketches and materialities of the studio often serve an epistemological purpose since it includes documented memories and investigations. Drawing and sketching can be seen as a 'form of individual inquiry' and research involving, theory, idea and action (Sullivan, 2007: p.240). This means that the importance of the studio as a knowledge space lies very much in it being a repository of materials, records and memories of artists' experiences being brought in from a world outside.

One of the artists says that she brings a notebook or sketchbook with her when she leaves her home or studio. She says 'I always have my research book with me. This one is a little bit too small though, I fill them up quickly. This one I started with two weeks ago, and now it is full already!' She calls them 'little exercise books' or 'research books' and they are very precious to her. She has kept all of them and they are all stored somewhere in her studio.

The artists I have talked with stressed the importance of itinerant experiences as impressions, immediate bodily sensation and sensuous experience in a variety of spaces they visit throughout their life and their occupation. Moreover, they described it as being important to transfer these sensations into notes and sketches; to obtain personal references and facts needed for future work; and to engage in classification work used to sort and organise these experiences. Through collecting and making sketchbook work, their sensations and experiences and journeys stop for a while and get transformed and transposed into an initial and referential form.

To collect or to construct references is a gathering process that is perhaps less about 'evidence' within their artistic practice compared to Latour's scientific case, but the sketches serve as what Latour (1999) calls 'silent witnesses' for certain and future claims and work. They do not work as proof, in the sense of truth and reliability, but the collected references might bring a continuity and coherence to the world that artists study. The sketches can be used as raw material in the making, but their function is also described to create and bring comfort to artists' work process. It is a practice through which their initial thoughts become physically manifested and therein they become 'visual evidence' of their work achievements and self-managed work. To collect, to store and to archive are practices then that are based on the embodied relations between the practitioner and the materialities of her making process. These artists often develop certain emotional attachments – such as comfort, care, obsession, enchantment and a sense of intrigue – with their collected impressions and sketches, which for example is expressed through the often overwhelming presence of kept physical collections and sketchbook archives in their studios. Through collecting and sketching, a certain curiosity is then developed for their material and project, which Sennett (2008) means is the first step towards knowing your material and eventually being able to make something.

By constructing sketches and references, artists' focus their attention and raise their awareness of their environments. The sketches function as recordings and representations, however they should not be seen to represent accurate records of something observed, as a mirror held up against their milieus, but rather what is being felt and imagined by the artists. Sketching can be seen as an interpreting practice of the surrounding environment. It involves the type of knowledge that Pocock (1981) suggests is developed and generated by 'observation' involving all the senses: looking that is not a matter of dis-embodied observation. One artist explains that what she sees, gets embodied and encoded and later expressed by, in this case, the workings of her hands. She explains it as a matter of

> perception more than seeing. It is not totally a visual thing. There is more to it. It is a more holistic perception. I am not thinking: oh what a nice composition or look at that particular angle of that street. But, I cannot really explain it, it is a deeper feeling. It is not entirely a logical or rational thing.

This sketchbook related observation and experience points to the geographical knowledge, that is their located, cultural, sensory and intellectual knowledge, in relation to how art gets made (Crouch and Toogood, 1999). Artists' making, observing and sketching are experienced through active hands, feet and minds – through bodily practices and interaction with the world in a direct mode. Artists' knowledge and investigations are not constructed through isolated meditation but through 'immersion of place and through bodily and intellectual exploration of that experience' (Crouch and Toogood, 1999: p.75). Their experiences are something they see, feel and think, hence observational sketching is related to a kind of practical theory or a mode of understanding that has less to do with abstract notions of discovery than with different possible ways in which they might relate themselves to their surroundings (Thrift, 2008: p.304).

Sketches can be seen as the outcome of individual interpretation and observation of the surrounding environment, and are also described to be somewhat prenotional: new ideas might emerge and take form through the sketching space itself. The skills involved in such drawing are described to be 'knowledge-in-potential', which is 'knowledge for and in becoming' (Rosenberg, 2008: p.124). Rosenberg writes about ideational drawing, which is a process at the same time as it is an artefact – it is mental and physical, bodily and cognitive. One of the artists explains her sketchbook to be 'a place where you write things by hand, just what you are doing now. And I really like that, because most of the time you don't. It is like writing a letter. It is a place where you see your own thoughts, originally written down.' Ideational drawing should not be seen as a way through which artists mainly commu-nicate but as a space for artists in which they can be by themselves and think: it is a 'denkraum'; a thinking space. (Rosenberg, 2008: p.123). The thinking space that drawing can create should be seen as a space where the artists' thoughts are represented but also presensed.

The sketchbook, the artists in this study suggested, can be described as a space where instructions, tasks and questions get formulated, and where their ideas and impressions often take their initial form. To draw and sketch is a step on the way towards an operationalization and materialization of initial ideas. Through the work of their hands, the artists were drawing as a practice through which they start to visualize an idea, to give form to thoughts in purposeful ways, to get closer to an understanding, to make sense of the world. However, sketches and sketchbooks are of course spaces where work is not only a matter of self-organization, reference and guidance, but an outcome of creativity in itself. The sketchbook was explained by one artist as her 'mini-studio'. To make a sketch in a sketchbook is already a creative and skilful outcome. Latour notes that 'the simple act of recording anything on paper is already an immense transformation that requires as much skill and just as much artifice as painting a landscape or setting up some elaborate biochemical reaction' (2007: p.136). Sketchbooks then can be understood on many levels but they are, I argue, vital parts of artists' making processes since they can be seen as primarily instructions and preparations in visual artists' individual projects. So what is the function and meaning of these artists' skilful practices of making sketches, preparations, plans and instructions in relation to their actual art making process?

Making sketches, making skills, making art

The artworks that I got to follow as part of this research project were not described as mainly idea-driven, but as emerging from complex patterns of finely controlled movements and repetitive practices of materials. It is through the relationship between the maker, a specific context, materialities but also lived action and practices that certain practical achievements will emerge. The form of artwork is, as stressed by Ingold (2001), generated in the course of the gradual unfolding of a sensuous engagement of the practitioner with the material he or she works with. To make art as well as developing new skills is a process constructed through the gradual attunement of movement and perception. In this process, each generation contributes to the next not by handing on a corpus of representations, or information in the strict sense, but rather by opening up opportunities for perception and action through providing the practitioner with the structures or platforms needed to be able to continue (Ingold, 2001: p.22). The making of art is dependent on skills and practice, and through the making of sketches and art, new practice and skills can emerge or get refined and modified, which open up new futures to the maker.

Through continuous practice and exercise, the practice can eventually become stored and embodied memories and skills, which were described to me to often come to dominate plans and initial ideas and sketches. In making processes directions and instructions are resources but as Ingold stresses 'they do not determine its course' (1996: p.180). One artist reflects on the relation between the preparation and making of art and between instruction and improvisation as a productive force in his studio based art practice. He sees preparation and sketching as practical rehearsals for his coming making practice. He explains his sketches to be

> essential, like rehearsing a move, or a shot in tennis. Then when it comes to the moment of making something you can do it without thinking. The state of mind, being relaxed, not thinking too hard about it, is the key, and continuous practice. ... There is not a lot I can say about the process, beyond the obvious, i.e. that it is improvised. Some visitors to my studio understand that, but others want some deeper explanation, or story, or theory.

This artist thinks that the force of practical improvisation and lived practice is particularly strong and at the core of his art practice (drawing) because of the

> free flow association of the line, that can loop, stray, zoom across, stop, describe, wander, zigzag, contain, decorate, and so on. It can be made to imply and suggest, whereas a splodge of colour is itself ... So a line can be relaxed, playful, suggest images as it unwinds. There is an affinity in this with the line of a melody, the line of a story, except that in drawing it is the process itself ... not laying out forms but making them up as it goes.

For one of the other artists, the relation between instruction and improvisation has become of particular interest in one of her projects: to learn how to make prosthetics. She explains how she has made preparations and instructions, such as sketches, photography and personal and practical advice in her digital sketchbooks. However, she explains how all these preparations were put aside when she was about to make something. Through preparatory photographs and sketches, 'you can see the potential and possibilities of the coming process, but they are not determining its course'. The preparatory work

> had already informed my knowledge about anatomical issues and hurdles and had provided me with a basic shape. But it outlived its use there, in this instance. This was partly because, what works on paper will not necessarily work in 3-D, but it was also very much about feel. My relationship with the piece changed when I began to sculpt it and I chose to make it in a less prescriptive way. The act of moulding and manipulating something with your hands is different to just looking at it. It is much more intuitive.

The relation between the material and their bodies, repetition and hands, involves a haptic and tacit knowing. Practice and information get transformed, embedded and embodied into tacit knowledge and over time practices might become so routinized that artists will not even reflect or think upon them anymore. It is the stored, bodily and taken for granted knowledge – the skilful relation between the body and the material – that then comes into focus. Artists' practical skills can be conceived of as being learnt through their repetitive but also varied practices around a particular task (Downey, 2005). An initial discomfort in practicing something new will be gradually effaced and the practitioner will eventually become less and less focused on the body. As Downey points out: 'a skill is finally and fully learned when something that was extrinsic grasped only through explicit rules or examples, now comes to pervade the own corporeality' (2005: p.27).

To be able to handle a particular technique or media opens up new professional and practical opportunities. Artistic practice and production can develop 'a mode of life that has more potentials available' (Deleuze, 1992 in Lea, 2009: p.473). Embodied skills and knowledge open up new ways of practising art as well as creating distinct characters of the artists. It is important to recognize that artists' intuition and improvisation does not come out of the blue, but, again, it is rather based on such accumulated skills. 'Thus, while improvisation, in its celebration of the moment, might appear on the surface to be a casual activity without preparation and consideration, it depends significantly upon great discipline, practice and experience' (Bain, 2003: p.312).

Through a making process and through repetition after repetition, the actual content of what is being made and learnt shifts. Learning and making is contextual and dependent on the context's changing character and the lived spontaneity, problems and events that are constantly emerging. Because of changing contexts, problems will continuously emerge which the learner has to be aware of and adjust to. One artist reflects on the relation between research and the sketches made on her computer and the actual lived experience in creating her art by hand. Even though she explains how she had a plan and image in her head of what it is she should be making, something else unfolds in practice. She means that the making process takes a new course whilst she is in the middle of it. She continues, 'you think you know what a kangaroo looks like, but when you start drawing, you don't. It is not until you make it, you realize the problems.'

Repetitive practice needs to be followed by self-monitoring as well as awareness and on-going readjustment, because of the constantly changing conditions of learning processes. Skilled practice is then not just the application of external force but involves qualities of care, judgement and dexterity. This implies that whatever artists do to things is grounded in an active perceptual involvement with what they watch and feel as they work. Schön (1983) writes about reflexive practice, which can have different characters. Reflexive practice involves reflection-in-action and reflection-on-action, processes through which a learner is able to 'surface and criticize the tacit understandings that have grown up around the repetitive experiences of a specialized practice, and make new sense of the situations of uncertainty or uniqueness which he may allow himself to experience' (Schön, 1983: p.61 in Barrett, 2007: p.118). The fact that preparation and lived practice, of for example the act of painting or sculpting, are different parts of the process is something that the artists are aware of, something they have learnt. In one of our conversations one artist tells me about a new idea of hers and she shows me a small prototype drawing in one of the notebooks, whilst stating: 'Oh here is my "height idea" I was talking about, but in reality it will not look anything like that.'

Sketches and sketchbooks can help visual artists to think around but also solve practical problems, as well as they support artists' learning and therefore to some extent help them anticipate what could occur in a making process. The creation of sketchbooks, collecting of objects and constructions of studio archives orient and structure their practice in the sense that it opens up certain possible paths of practice and closes others down. Practices and practical knowing, as Ingold suggests, builds on knowing and following materials and substances of the making: 'As practitioners, the builder, the gardener, the cook, the alchemist and the painter are not so much imposing form on matter as bringing together diverse materials and combining or redirecting their flow in the anticipation of what might emerge' (2010: p.94). The productive force of studio based making partly lies in the particular collection of experiences, resources, materials or 'substances', rather similar to that of the alchemical laboratory.

Conclusion

By focusing on the function and meaning of the sketchbook as a knowledge making device this chapter offers glimpses into visual artists' skilled and studio based making. To make sketches and fill out sketchbooks is a creative act in itself, but in this chapter sketches and sketchbooks have also been suggested to be vital resources in artists' making and learning. Through sketching artists engage in individual inquiries, research and the making of reference, self-directed instructions and preparations. The sketchbook is described as a companion to artists' itinerant practice outside the studio, it helps focus their attention and raise their awareness of their environments; it is a device that supports artists' development of their geographical knowledge. Artists' knowledge and investigations of the world through sketching are not constructed through isolated meditation but through the immersion of place that involves active engagement through embodied observation and interpretation of their contexts. Sketchbooks are a context where new ideas or thoughts are represented as well as they are a space where ideas are invented and presensed. Through making sketches, artists can think with and through drawing to make discoveries, to find new possibilities that give course to ideas and help them take its eventual form. The skills and knowledge involved in such sketching have been described with Rosenberg's idea of 'knowledge-in-potential', which is 'knowledge for and in becoming'. Through the continuous practice and experience of making sketches artists develop and accumulate skills that get embodied and performed in their making of art. Sketches open up production possibilities since they help artists solve practical problems and challenges that they anticipate could emerge in a making process.

Practices and devices such as sketching and sketchbooks create clear links between different spaces of artistic practice. In particular, there is a clear set of links made between spaces of artistic making – principally the studio – and spaces of exploration, learning and research. When it comes to the actual making of art in the studio, however, sketches, references and instructions are present but often put to one side. The studio is a space where artists' ideas materialize and take form through skilful and finely controlled and repetitive movements and tacit and sensuous engagement with materials. In this process the conditions of making processes are continuously changing. Tacit knowledge and making is dependent on this lived spontaneity of making and the changing character of the maker's context. Because of changing contexts, problems will continuously emerge which the learner has got to adjust and importantly stay attuned to.

To make art is based on skilful practices that are relational in character. Artists generate and apply their skills through actively engaging with specific work devices, surroundings and contexts. In this study it has been suggested that artists' making is related to the studio as a workspace and workshop, where their making feeds from collected material, sketches, references, documentation and research, as well as from their lived action and practice in the studio. The generated knowledge and skills of the studios are properties of the whole human-organism-person that have emerged through the history of her involvement with that environment. The collected sketches, memories and references that form the materialities of the studio are fundamental resources in the making of art.

References

Alpers, S. (2010) The View from the Studio. In Jacob, M. J. and Grabner, M. (eds) *The Studio Reader: On the Space of Artists*. Chicago: University of Chicago Press.

Amirsadeghi, H. (2012) *Sanctuary: Britain's Artists and Their Studios*. London: Thames & Hudson.

Bain, A. (2003) Constructing Contemporary Artistic Identities in Toronto Neighbourhoods. *The Canadian Geographer*, 47(3), 303–317.

Bain, A. (2004) Female Artistic Identity in Place: The Studio. *Social and Cultural Geography*, 5(2), 171–193.

Bain, A. (2005) Constructing an Artistic Identity. *Work, Employment and Society*, 19(1), 25–46.

Barrett, E. (2007) Experiential Learning in Practice as Research: Context, Method, Knowledge. *Journal of Visual Arts Practice*, 6(2), 115–124.

Buren, D. (1970/71) The Function of the Studio. In '*The Studio*' exhibition catalogue. Dublin: Dublin City Gallery.

Buren, D. (2007) The Function of the Studio (Revisited): A Conversation between Daniel Buren and the Curators/Editors Jens Hoffmann, Christina Kennedy and Georgina Jackson. In '*The Studio*' exhibition catalogue. Dublin: Dublin City Gallery The Hugh Lane.

Chare, N. (2006) Passages to Paint: Francis Bacon's Studio Practice. *Parallax*, 12(4), 83–98.

Crouch, D. and Toogood, M. (1999) Everyday Abstraction: Geographical Knowledge in the Art of Peter Lanyon. *Cultural Geographies*, 6(1), 72–89.

Daniels, S. (2011) The Art Studio. In Agnew, J. and Livingstone, D. (eds) *The SAGE Handbook of Geographical Knowledge*. London: Sage.

Davidts, D. and Paice, K. (eds) (2009) *The Fall of the Studio: Artists at Work*. Amsterdam: Valiz.

Deleuze, G. (1992) Postscript on the Societies of Control, *October59*, 3–7.

Downey, G. (2005) *Learning Capoeira: Lessons in Cunning from an Afro-Brazilian Art*. Oxford: Oxford University Press.

Farías, G. and Wilkie, A. (2016) *Studio Studies: Operations, Topologies and Displacements*. New York: Routledge.

Gill, R. and Pratt, A. (2008) In the Social Factory Immaterial Labour, Precariousness and Cultural Work. *Theory, Culture & Society*, 25(7–8), 1–30.

Graw, I. (2003) Atelier. Raum ohne Zeit; Vorwort. Texte Zur Kunst, 13(49). Cited in: Davidts, D. and Paice, K. (eds) (2009) *The Fall of the Studio: Artists at Work*. Amsterdam: Valiz.

Grabner, M. (2010) Introduction. In Jacob, M.J. and Grabner, M. (eds) *The Studio Reader: On the Space of Artists*. Chicago: University of Chicago Press.

Hawkins, H. (2010) Turn your Trash into… Rubbish, Art and Politics. Richard Wentworth's Geographical Imagination . *Social and Cultural Geography*, 11(8), 805–827.

Hawkins, H. (2012) Geography and Art. An Expanding Field Site, the Body and Practice. *Progress in Human Geography*, 37(1), 52–71.

Hughes, A. (1990) The Cave and the Stithy: Artists' Studios and Intellectual Property in Early Modern Europe. *Oxford Art Journal*, 13(1), 34–48.

Ingold, T. (1996) Situating Action V: The History and Evolution of Bodily Skills. *Ecological Psychology*, 8(2), 171–182.

Ingold, T. (2001) Beyond Art and Technology: The Anthropology of Skill. In Schiffer, M. B. (ed.) *Anthropological Perspectives on Technology*. Amerind Foundation New Worlds Studies Series. Albuquerque: University of New Mexico Press.

Ingold, T. (2010) The Textility of Making. *Cambridge Journal of Economics*, 34(1), 91–102.

Jacob, M. J. (2010) Preface. In Jacob, M. J. and Grabner, M. (eds) *The Studio Reader: On the Space of Artists*. Chicago: University of Chicago Press.

Jacob, M. J. and Grabner, M. (eds) (2010) *The Studio Reader: On the Space of Artists*. Chicago: University of Chicago Press.

Latour, B. (1999) *Pandora's Hope. Essays on the Reality of Science Studies*. Cambridge, Massachusetts: Harvard University Press.

Latour, B. (2007) *Reassembling the Social. An Introduction to Actor-Network-Theory*. Oxford: Oxford University Press.

Latour, B. and Woolgar, S. (1979) *Laboratory Life*. Princeton: Princeton University Press.

Lea, J. (2009) Becoming Skilled: The Cultural and Corporeal Geographies of Teaching and Learning Thai Yoga Massage. *Geoforum*, 40(3), 465–474.

Mackenzie, A. F. D. (2006) 'Against the tide': Placing Visual Art in the Highlands and Islands. *Social and Cultural Geography*, 7(6), 965–985.

Madestrand, B. (2012) *Konstnärer och deras ateljéer*. Stockholm: Bladh by Bladh.

Major, G. (2012) *Vis Loci: Geographies of the Artist's Studio Exhibition Catalogue*. London: Squid & Tabernacle.

Markusen, A. (2006) Urban Development and the Politics of a Creative Class: Evidence from a Study of Artists. *Environment and Planning A*, 38(10), 1921–1940.

Morris, N. J. and Cant, S. G. (2006) Engaging with Place: Artists, Site Specificity and the Hebden Bridge Sculpture Trail. *Social and Cultural Geography*, 7(6), 863–888.

Pocock, D. C. D. (1981) Sight and Knowledge. *Transactions of the Institute of British Geographers*, 6(4), 385–393.

Roberts, J. (2007) *The Intangibilities of Form Skill and Deskilling in Art after the Readymade.* New York: Verso.

Rogers, A. (2011) Butterfly Takes Flight: The Translocal Circulation of Creative Practice. *Social and Cultural Geography*, 12(7), 663–683.

Rosenberg, T. (2008) New Beginnings and Monstrous Births: Notes Towards an Appreciation of Ideational Drawing. In Garner, S. (ed.) *Writing on Drawing: Essays on Drawing Practice and Research.* Bristol: Intellect Books.

Ross, A. (2008) The New Geography of Work Power to the Precarious? *Theory, Culture & Society*, 25(7–8), 31–49.

Sachs Olsen, C. and Hawkins, H. (2016) Archiving an Urban Exploration – MR NICE GUY, Cooking Oil Drums, Sterile Blister Packs and Uncanny Bikinis. *Cultural Geographies*, 23(3), 531–543.

Schön, D. A. (1983) *The Reflective Practitioner.* New York: Basic Books.

Seijdel, J. (2009) Editorial. In *Open 17: A Precarious Existence. Vulnerability in the Public Domain.* Rotterdam: NAi Publishers SKOR.

Sennett, R. (2008) *The Craftsman.* New Haven: Yale University Press.

Sjöholm, J. and Pasquinelli, C. (2014) Artist Brand Building: Towards a Spatial Perspective. *Arts Marketing: An International Journal*, 4(1/2), 10–24.

Sullivan, G. (2007) Creativity as Research Practice in the Visual Arts. In Bresler, L. (ed.) *International Handbook of Research in Arts Education.* Dordrecht: Springer.

Thrift, N. (2008) *Non-representational Theory. Space, Affects, Politics.* New York: Routledge.

Wainwright, L. (2010) Foreword. In Jacob, M. J. and Grabner, M. (eds) *The Studio Reader: On the Space of Artists.* Chicago: University of Chicago Press.

Waterfield, G. (2009) *The Artist's Studio*, London: Hogarth Press.

3 Moonraking in Slaithwaite

Making lanterns, making place

Tim Edensor

Introduction

In the West Yorkshire mill town of Slaithwaite, on the evening of 21 February 2015, a large, illuminated paper lantern forged into the shape of a moon was hauled out of the canal by an adjacent crane. Thereafter, it was installed at the head of a 400 strong procession in which most participants were carrying candle-lit paper lanterns. This procession, the culmination of Slaithwaite's biannual *Moonraking* festival, celebrated its 30th anniversary.

In this chapter, I discuss the different kinds of making that have been involved in *Moonraking* since its inception, particularly focusing upon the crafting of the lanterns by local participants and artists, and the place-making and event-making qualities of the occasion that are sustained by myth-making, playfulness and conviviality, and the production of a potent atmosphere. I foreground the essential inclusiveness that succours the ongoing creativities that emerge as well as the ways in which the festival's duration fosters a deep experience of making in different ways.

Moonraking was devised and initiated to celebrate a mythical event involving the smuggling of illicit alcohol that took place along the Huddersfield Narrow Canal that passes through the town, a story that seems to capture the town's anti-authoritarian spirit and local cunning. One night, at the turn of the 19th century, several Slaithwaite men returned to the canal to retrieve barrels of liquor that had been delivered earlier in the day by barge and hidden underneath one of the bridges. As they were fishing out the barrels, two excise men caught them in the act and loudly demanded to know what they were up to. Pretending to be thoroughly intoxicated, one of the men replied that the moon had fallen out of the sky and they were attempting to rake it out of the canal, tricking the officials into thinking they were merely drunken fools. They got away with it.

Thus, the festival commences with the hauling out of the large moon-shaped lantern from a barge moored in the canal and its positioning at the head of the parade. Behind the moon are an assembled throng accompanied by several bands garbed in colourful illuminated decorations, who for

Figure 3.1 Lanterns fashioned into landmarks, 2015 *Moonraking* festival, Slaithwaite
Photograph by Tim Edensor

50 minutes or so make their way around a route that incorporates different parts of the town. Most of these participants hold aloft lanterns composed out of willow twigs and paper that they have created during workshops staged in the week prior to the event. The lanterns are devised according to a particular bi-annual theme. In 2013, this was 'time', and lanterns were wrought into forms resembling grandfather clocks, egg timers, pyramids and clock towers, whereas in 2015, a 'landmark' motif encouraged participants to make lanterns in the shape of the Eiffel Tower, the Taj Mahal and the London Eye, amongst other iconic structures. The procession culminates with the return to the starting point, and with fireworks and dancing in the streets.

It is important to understand the political and historical context of *Moonraking's* instigation in 1985. Its founders, still organizing the festival, are Gill Bond and Andy Burton, who after working as freelance community outreach workers with renowned radical theatre group Welfare State International, and influenced by their philosophy, set up Satellite Arts, which remains located in Slaithwaite. The aim was to deepen community engagement and develop diverse skills and creativity within communities that could act as satellites for further emergent practices.

Welfare State International (WSI), founded in 1968, developed their distinctive approach to community arts in the intensified political aftermath of the revolutionary events of the 1960s. Yet rather than focusing upon direct political action, inspired by radical North American groups, WSI emphasized the importance of developing the resources of imagination, embedded in a commitment to co-producing experimental, collaborative and inclusive ceremonial and theatrical events. The site-specificity of these temporary multi-media performances and processions, that typically involved the use of fire, ice, sound and light, was integral to their political motivation to involve people unused to or uninterested in more formal drama (Whiteley, 2010). In staging workshops and festivals in schools, housing estates and workplaces, a collaborative politics of creativity was prioritised, nurturing a sense of ownership and recognition.

According to Gillian Whiteley (2010), a key ingredient in the development of these creative events was myth. Conceived as potentially liberating and exemplary, notably through reiterative performance, this use confounded understandings that regard myth as inherently conservative. By soliciting a mythic context through which current circumstances can be temporarily transcended, imaginatively, socially and sensually, WSI aimed to solicit an awareness of the potential for enhancing collectivity and expression. For Michael White (1988), formerly a member of WSI, such events could act as a kind of homecoming in which vital links between the domestic, the social, and the mythic might be re-established. Moreover, as I discuss in greater detail below, over time the affective charge of reiterative events such as *Moonraking* can contribute to strengthening both a sense of community and developing a creative criticality.

The origins of *Moonraking* were motivated following a visit to Japan by WSI. During the trip, the potential of the lantern fabricated out of willow sticks was identified, since it could be made into innumerable shapes. Such lanterns could be constructed without a great deal of expertise or prior knowledge but could also be fashioned into innovative, artistic forms. Accordingly, as a creative practice, lantern making afforded an inclusivity amongst makers and offered to reconnect people with the means of cultural production through participatory artistic activity. The affordances of these lanterns, light in weight and safe, encouraged the staging of a parade in which they could be carried and displayed.

The heart of *Moonraking* remains the fabrication of lanterns. In recent years, lantern parades have spread across the UK and beyond and can be conceived as part of the wider creative deployment of light for festive and artistic purposes that includes the design of light sculptures as iconic place-making installations (see Edensor, 2015a, b; Hawkins, 2015; Papadaki, 2015; Sloan, 2015) and large national ceremonies such as the opening ceremonies of the Olympic Games and Australia's Anzac Day (Sumartojo, 2015). A plethora of techniques has emerged as light festivals have taken increasingly varied characteristics, ranging from huge spectacular events such as Lyon's *Fête des Lumières* and Sydney's *Vivid* to the small scale lantern parades typified by *Moonraking* (Edensor, 2015, 2017).

Though light festivals might appear inclusive in generating public engagement and enjoyment, they have been critiqued as synonymous with 'festivalization' or 'eventification', instrumental strategies conceived as central to the promotion of the 'creative city'. In such critiques, creativity is a key notion firmly welded to highly valorised expressions of neo-liberal regeneration, and accordingly, is hyperbolically deployed to summon up a 'creative class', the 'creative industries' or 'creative cities' (Florida, 2005). In culture-led regeneration strategies, heritage, art and leisure are promoted as 'experiences' to be consumed in 'tourist resort areas, theme parks, redeveloped water-fronts, trade expositions, mega-events, shopping complexes, festival markets, art shows and galleries, opera halls and museums' (Gotham, 2005: 227). As cities vie for inward investment, these place-enhancing culture-led strategies offer evidence of a 'vibrant' location that will attract tourists, shoppers, investors and potential middle-class residents.

It is certainly the case that since the 1980s, there has been a huge rise in the number of urban arts festivals that have become an essential element in place-making strategies and 'a mainstay of urban tourism and urban policy-making' (Quinn, 2010: 266). While initially, such festivals were large in scale, Doreen Jakob contends that cities have subsequently developed a much more extensive process of 'eventification', which has 'infiltrated urban and economic development on a much smaller scale'. She alleges that such events incorporate the 'deliberate organization of a heightened emotional and aesthetic experience at a designated time and space' (2012: 448). This resonates with Pine and Gilmore's (1999) insistence that places must stage experiences to add sufficient value to their economy, thereby producing public spaces and events as 'a spectacle and transformed into an aestheticized place of consumption' (Jakob, 2012: 449). Disguising economic disinvestment and growing social inequality, festivals are part of a broader, Darwinian inter-urban competition. Critics suggest that instead of generating creativity, they tend to produce a host of aesthetic conventions, designscapes, events and artistic forms, described by Guy Julier as a collection of 'brand design, architecture, urban planning, events and exhibitions' (2005: 874), qualities promoted by the aforementioned international creative class. Because only particular aesthetic and artistic conventions predominate in this redistributing of the sensible (Rancière, 2009), more challenging and critical creative expressions are squeezed out. An avoidance of promoting anything too controversial or challenging in culture-led, place-branding strategies may consequently oblige artists to produce bland works that attract large crowds and satisfy key economic stakeholders. Accordingly, as Kevin Fox Gotham contends, 'whereas the appeal of local celebrations is the opportunity to see something different, celebrations that are redesigned to attract tourists seem more and more alike' (2005: 234). Such critiques infer that generic spectacles eventuate, as 'cultural substance becomes replaced with cultural spectacle', and 'increasing homogeneity and declining creativity' result as 'cultural strategies fail to connect with the specificities of the places within which they are located' (Quinn, 2010: 271–2).

These critiques chime with earlier claims that we now live in a 'society of the spectacle' in which spectators passively behold seductive, extravagant displays organized by capital and the state that have replaced 'authentic' life (Debord, 1994). Characterized as a new stage in the development of capitalism, the society of the spectacle shifts from manufacturing commodities to producing images, and is dominated by advertising, mass media and other culture industries while simultaneously stimulating consumer desires. As social relations become increasingly mediated by commodified images, estrangement become pervasive. Such spectacles bedazzle people, limiting their ability to perceive the 'real' conditions that underlie their enslavement as workers and consumers. Distracted by fantasies that pacify and seduce, spectators' scope for critical thought is diminished.

While there is no doubt that these pessimistic accounts possess some contemporary salience, their sweeping claims about notions of passive spectatorship surely deny the creativity and reflexivity of an active audience (Barker, 1999). The capacity of onlookers to interpret what they see in numerous ways, irrespective of the meanings encoded into the forms upon which they gaze, is denied (Hall, 2001). Crucially, they also neglect the more productive effects that festivals can promote, for even at very large events, there is scope for ingenuity in manufacturing and consuming displays, in place-making, and in soliciting forms of conviviality and playfulness. Perhaps at smaller festivals like *Moonraking*, such progressive effects are more evident, for there is little compulsion to prioritize promotional and commercial concerns, as I now discuss in greater detail.

Vernacular and artistic lantern making at *Moonraking*

David Picard and Mike Robinson emphasize that 'festivals form privileged arenas of cultural creativity' (2006: 14) in which the novel, the prototypical and the experimental enter the field. They are apt to serve as occasions on which constellations of artistic creativity, technical expertise and public design emerge, offering opportunities to experiment, test potentially useful technologies, and share ideas and skills. At festivals of illumination, groups of designers, artists, light technicians, computer technicians and engineers may try things out in situ. Besides the coming together of professionals to share ideas, techniques and know-how, experienced locals also contribute to the dynamic and fluid qualities of the event. Such processes require that the festive environment be inclusive and open-minded so as to 'offer different social groups open access and a variety of opportunities to share experiences and encounter the new' (Schulte-Römer, 2013: 152). A guiding imperative for *Moonraking* has been to expand the possibilities for being creative in the community, encouraging collective involvement in the production of the event.

This has primarily depended upon the workshops that are organized in the week leading up to the event to engage locals – adults and schoolchildren – in the making of the willow and paper lanterns central to the parade. The participatory impulse of the workshops is to accommodate the different skill levels and experience in making possessed by the diverse participants. The outcome is to manufacture the range of diverse lanterns that make their way around the parade, a motley collection of exhibits that testify to different levels of expertise and effort.

Crucially then, the parade is a celebration of the vernacular creativities (Edensor et al., 2009) that have become sedimented amongst those residents of Slaithwaite who have participated over the years and that are expressed as new participants become involved. In considering these expressions, the discussion offered by Elizabeth Hallam and Tim Ingold is instructive. They argue that creativity should be conceived as an improvisational quality across all forms of cultural activity that requires people to adapt to particular circumstances. Even in apparently repetitious practices such as the annual workshops held in Slaithwaite, regeneration takes place under circumstances which are invariably different. Accordingly, although the designs of many of the lanterns are inventive and ingenious, it is inappropriate to only regard these constructions as exemplifying creativity. For ideas that creativity is consistently innovative and concerns the production of novelty ignores how cultural practice 'entails a complex and ongoing alignment of observation of the model with action in the world' (Hallam and Ingold, 2007: 5). Hallam and Ingold thus insist that there 'is creativity even and especially in the maintenance of an established tradition... (for) traditions have to be worked to be sustained' (ibid.: 5–6), and moreover, such reiterative processes often stimulate improvisation.

Importantly, Hallam and Ingold decentre notions that creativity is the property of a gifted individual or genius. Instead, it is relational, involving 'persons in those mutually constitutive relationships through which, as they grow older together, they continually participate in each other's coming-into-being' (ibid.: 6). The idea that the uniquely creative person can somehow disentangle themselves from the social (including the non-human) world is dismissed since, as 'it mingles with the world, the (individual) mind's creativity is inseparable from that of the total matrix of relations in which it is embedded and into which it extends' (ibid.: 9). Implicitly then, creativity is social and sociable, culturally specific and communally produced. The sequence of events through which *Moonraking* comes into being exemplifies this creative process.

Many of those who have repeatedly fabricated lanterns over the years have developed an embodied skill in deciding upon the form, weight and structure of the piece. A cognitive and sensuous practice of making becomes sedimented in the accumulation of knowledge about the constitutive materials and techniques (Miller, 2016). The lanterns are not merely inert sculptures, and this embodied skill is supplemented by an understanding about how the lantern will be wielded during the procession and a sense of how it will appear as it mingles with the other lanterns, thus connecting the process of making with the performance and experience of place.

A shared aesthetic and expertise has been maintained by the consensual decision among organizers and participants that the specific glow of the lanterns must be produced by candlelight, which bestows a particular luminescent quality that is radiated by hundreds of lit creations. This aesthetic coherence is also preserved by an insistence that the lanterns should be made out of white paper (although used in the past, colour is largely rejected), policies that produce an evenness of effect despite the variable qualities of the individual designs. The suffusion of dark space with a sea of bobbing, diffuse white lights is generated by a collective production that maximizes inclusion.

The promotion on local participation in lantern-making has generated a broader enthusiasm for cultures of making. This is exemplified by how, since 2013, the *Moonraking* festival has been augmented by a 'Handmade Trail' where local and regional artists and craft-makers showcase and sell their work to visitors at pop up markets.

However, this inclusive insistence on showcasing the vernacular creativities of variously skilled locals has not eclipsed the important role played by professional artists and technicians in contributing to *Moonraking*. For from the first staging of the event, the all-important moon has been fabricated by a skilled lantern maker. Over the years, the locally wrought lanterns have been supplemented by a plenitude of other illuminated installations. In 2015, for instance, the festival gained local and national funding to organize three commissions that focused on the 'art' of lantern-making, inviting artists to engage with community groups to create large set-piece pieces that were subsequently situated along the route. A large lantern, *Gnome of the North*, was based on Anthony Gormley's renowned Tyneside colossus, the *Angel of the North* and designed by Karima Ellis, a lantern-maker whose practice has been developed over 25 years of *Moonraking*. Co-founder Andy Burton, produced the substantial *Moonhenge*, formed by lanterns arranged into a scaled down version of Stonehenge. Finally, the *Lantern House* was also produced by two artists with long-standing associations with the festival. Collectively known as PaBoom, they produced a garden shed-sized structure that by contrast deploys colour, and is adorned with paper-cuts to create shadows and static lights that illuminate the house.

Other artist acquaintances regularly contribute to *Moonraking*, underpinning how the festival has become an enduring occasion to which makers can tailor their designs. This allows for a regular involvement that

Figure 3.2 The 'Angel of the North' Lantern, Karima Ellis, 2015 *Moonraking* festival,
       Slaithwaite
Photograph by Tim Edensor

Figure 3.3 'Moonhenge', Andy Burton, 2015 *Moonraking* festival, Slaithwaite
Photograph by Tim Edensor

Figure 3.4 'Lantern House', PaBoom, 2015 *Moonraking* festival, Slaithwaite
Photograph by Tim Edensor

fosters the potential to develop skills over the week and year on year. Yet
the event acts as a creative hub that exceeds its calendrical occurrence,
inspiring and generating work that is also displayed at other events across
the north of England and beyond, as a growing network of festivals continue to emerge.

Distinctions between the vernacular and more established creative practices
and subjects have become blurred during the 30 years in which *Moonraking*
has been staged, as exemplified by how creative engagement for several participants has culminated in their undertaking careers in the arts. Yet the many
local participants who do not take such a route also benefit from being
involved in a project that has endured for many years and relies on an accumulating resource of situated know-how and creativity. Besides contributing
to the ongoing evolution of the specific character of the festival, it has also
encouraged a dynamism wherein new creations add to the compilation of
diverse artistic forms over time. In future events, it is envisaged that this
continually emerging scenario will incorporate mechanized and animated
lantern forms and an increase in the numbers of static lanterns displayed in
the windows of households passed during the parade.

The history of those who have been and continue to be engaged in *Moonraking* involves their adoption of various roles and participatory practices,
including lantern-making, workshop organization, administration, stewarding

the parade and management that has developed a collective accrual of expertise and skill. These biographies contribute to a wider sense of ownership of the event and have fostered a powerful sense of community through the forging of connections within and beyond Slaithwaite. Like the pot-making practices discussed by Alex Miller, lantern-making in Slaithwaite can 'express and embody individuals' identities while at the same time emplacing them in wider networks of relatedness' (2016: 5).

Place-making and event-making at *Moonraking*

In considering the making processes of *Moonraking* it is essential to acknowledge how a sense of occasion is produced and this combines with the place-making qualities of the event. As with most festivals, *Moonraking* is of short duration, but in the build up to the parade, filled with preparation and anticipation, and on the night of the procession, normative quotidian rhythms are temporarily banished. In marking out a transitory departure from everyday routines, festivals constitute a period during which ordinary conventions and social interactions can be suspended and there is greater potential for participants to engage in expressive, creative and improvisational behaviour (Edensor, 2010). Festivals such as *Moonraking* are ephemeral and only briefly colonize space, lacking the capacity to permanently transform the sedimented meaning and practice of place. Nevertheless, such occasions do offer opportunities for practising, representing and apprehending place in ways at variance to habitual experience. They propose that everyday life and community might be experienced otherwise, and space that usually serves as a location for everyday work, commerce, travel and mundane tasks is transformed into a convivial playground.

In Slaithwaite, the meaning, feeling and function of small town space is altered as participants suspend self-consciousness and instrumental dispositions. Picard and Robinson capture how festivals may produce situations in which the usual 'spatial functionalities become hidden and forgotten, signs become meaningless, directions reverse, boundaries cease to bound and the mundane is decorated and disguised and overtaken by different rituals and practices' (2006: 11). This resonates with the Situationist strategies of *detournement* through which the habitual and the taken-for-granted are made strange (Knabb, 2006). Indeed, Welfare State International were highly influenced by Situationist ideals of play, self-expression and creativity but grounded their cultural practice in highly participatory endeavours. Thus rather than an obscure practice for select initiates, many festivals democratize the critical potential sparked by making the ordinary weird. Defamiliarization of place and space through illumination and parade is thus integral to *Moonraking*, with its efflorescence of vernacular surrealism.

Moonraking demonstrates how light possesses a particularly rich capacity to defamiliarize familiar places, transforming that which is well known into an uncanny realm. This was given impetus as illumination became part of the broader modern colonization of aesthetic experience into ever more spaces, producing an abundance of 'new perceptual pleasures' (Böhme, 2010: 29). Along with telephony, film and photography, electric light has added to what Collins and Jervis term the phantasmagorical city, replete with 'the shadowy hauntings of the fleeting and insubstantial', producing defamiliarization, uncertainty and fascination, constitutive aspects of modern experience (2008: 1). Through the deployment of lighting in fairgrounds, theatres, shop windows and advertising displays, amongst other venues, nocturnal cityscapes have emerged that challenge perception and attract vision towards the fantastic.

Such encounters with the dreamlike, surreal and strange foster what Martyn Evans (2012) terms 'wonder', which he characterizes as an 'altered, compellingly intensified attention' that saturates experience in the moment through which the world is made *newly-present* to us. Under conditions of such intensified perception, we attend more intently to the previously over-looked or undervalued. The challenging experiences of seeing familiar space in novel ways also temporarily transforms place by communicating amazement, pleasure and amusement, contributing to what Fisher and Drobnick term 'the *nocturnal carnivalesque*' which 'defamiliarizes the city as well as opens it up to alternative interpretations and possibilities' (2012: 36). At Slaithwaite, the bizarre and fantastic lantern forms carried in the parade and the particular glow bestowed by white paper and simple candles charge the event with a distinctive aesthetics and a sense of unreality. These qualities invest the occasion with festive alterity and compound its capacity to enchant time and space.

In addition to the ways in which *Moonraking* defamiliarizes the setting in which it takes place, another festive quality continuously reproduced is the license for playful interactivity that pervades Slaithwaite's streets. Among participants and observers, the event encourages the widespread adoption of expressive, creative practices and a convivial disposition towards others that signals a temporary switch towards a more intensely affective and sensual engagement with place and its inhabitants. *Moonraking* thus exemplifies how festivals coax people into creative expression and extrovert display that claims space as a venue for fun, frivolity, good humour and collectivity (Stevens, 2007), in contradistinction to the critiques of festivals discussed above that suggest that they serve as occasions when onlookers are hypnotized by spectacle. While such practices certainly do not constitute a radical political resistance to the established order, they possess other qualities: as Woodyer claims, 'the politics of playing are primarily bound up in experiencing vitality rather than strategic oppositional endeavour' (2012: 318). Such playful festive engagement can be conceived as a form of (re)making place, a practice through which everyday space is reterritorialized in ways that re-enchant a sense of belonging and shared habitation

Relatedly, festivals such as *Moonraking* are occasions at which a key element of their production and experience is *atmosphere*, a festive atmosphere that is made out of a medley of elements, including the playful enthusiasm of participants, anticipation, defamiliarization, illumination, and the musical contributions of the bands. Atmospheres are not formed out of one element but continuously emerge out of an amalgam of forces, affects and happenings. These constituents are distributed across space and circulate between people, spaces and things, since atmospheres are relational phenomena that enrol different configurations of objects, technologies and (human and non-human) bodies in their ongoing emergence (Stewart, 2011: 445). Rather than constituting a durable condition, atmosphere flows as a sequence of events and sensations, successively provoking immersion, engagement, distraction and attraction. During *Moonraking*, the potent atmosphere is particularly generated by the ways in which people respond to and communicate their feelings through movements, gestures, voices and faces. Crucially, because the festival is important to the town, these affective and emotional expressions are stoked up by an anticipation amongst participants for the events that are to unfold, an anticipation borne out of previous memories, lantern preparation and daily conversation. A shared emotional and affective relationship with the occasion has been woven into family and communal histories (Edensor, 2012). And the significance of *Moonraking* is re-made by the powerful atmosphere and its co-production by residents' anticipatory dispositions and subsequent participation.

In addition, festive light design may conjure up powerful references to histories and qualities that foment 'a sense of belonging to and understanding of places, giving new meanings so that territories are repossessed' (Alves, 2007: 1259). This is evidently the case during *Moonraking*, where the parade and the myth upon which the festival is founded underpin the reiteration of place-specific qualities. For the myth and its bi-annual evocation posits Slaithwaite as a town at which the rules and ethos of the powerful are not necessarily followed, may be subverted and bypassed in cunning, improvisational ways. The independent spirit, though romanticized through the retelling and restaging of the defiant act of 19th century residents, offers an exemplary quality that can be deployed metaphorically towards contemporary impositions of bureaucracy and commerce. The tale and performance thus mark the town as distinctive, and this is underpinned by the uniqueness of *Moonraking* itself.

Equally evidently, place-making is forged by the parade, a communal event co-produced by organizers, those walking and carrying lanterns, and the many bystanders that line the route to watch. The event takes possession of streets that usually privilege vehicular traffic and animates usually quiet spaces with noise, light and movement. The ritualized, theatrical beginning of *Moonraking*, with its costumed figures and ceremonial hoiking of the moon out of the canal, builds anticipation, and as the moon-lantern takes its place at the front, the parade commences with jollity and enthusiasm, and subsequently charts a path through diverse areas of the town. Thereafter, it moves along a gloomier, quieter path next to the formerly disused canal where the white lanterns reflect in its still water, and then passes underneath the railway, entering a deep, dark tunnel in which the lanterns cast light and shadows across its curved ceiling and the participants create an uproar that echoes through the confined space. Following this, the parade enters streets lined by onlookers as other residents shout out from the upstairs windows of adjacent houses, engaging in frivolous banter with the lantern wielders. The route then moves past a tree-lined area before returning to the starting point.

Accordingly, *Moonraking*, like other festivals, offers an opportunity when 'individuals and groups can discursively manifest their visions of the world and create meaningful frameworks of their being together', transmitting identities, cultural practices and values to locals and outsiders (Picard and Robinson, 2006: 12). Instead of grand and sober ceremonial parade, *Moonraking* is a creative, improvisational, convivial and celebratory form of commemoration and place-making replete with sensual and playful pleasures that reinscribes a route around the discrete parts of the town. And crucially, rather than the performance of reiterative ritual borne out of the reified, authoritatively delineated heritage of place, the parade celebrates some of the forgotten figures of history, honouring an anti-authoritarian incident in which the strictures imposed by authority were cunningly sidestepped.

Sidestepping neo-liberal cultural policies

In this chapter, I have explored how *Moonraking* has continuously changed in many ways. Though the festival's myth, rituals and parade route have become stabilized, and the focus remains on the production and display of paper and willow lanterns, the event has adapted and mutated over the years. This setting has encouraged new participants, innovations, artistic works and associations but *Moonraking* has retained an inclusiveness that continues to offer fertile conditions in which forms of making endure and emerge. For all these participants, as Carr and Gibson submit, making is 'a material conversation – a physical provocation and a response, iterated over and again, working with the material to understand its capacities, analyze error and make adjustments' (2015: 7). Besides the crafting of lanterns, I have foregrounded the making of place, event and atmosphere as integral to *Moonraking's* persistence in Slaithwaite. The 30 year duration of the festival has allowed artistic and vernacular creativities to develop and promoted the ongoing formation of a network of diverse lantern-makers and other participants. This is celebrated in a publication produced on the 21st anniversary that records the evolution of the festival, its organizational changes and artistic shifts and key events, and honours those who have participated over the years (Bond, Burton and Ellis, 2006).

This case reinforces the valuable contribution that festivals can contribute to fostering a sense of community and promoting creativity, qualities acknowledged by Arts Council England, who have come to regard outdoor performance and street arts as compelling elements in cultural policy and planning, and in the enhancement of place. As Micklem declares, by 2006, carnival was recognised as 'an extraordinarily potent force with regeneration, social inclusion, participation and tourism' (2006: 4). *Moonraking* also underscores how light festivals have the potential to defamiliarize and enchant space with an aesthetic alterity, and this is expressed by the vernacular surrealism presented by a sea of bobbing white lanterns forged into timepieces or landmarks, as well as the inventive artistic installations that accompany them.

This recognition of the value of festivals belongs to the broader process through which urban policymakers have foregrounded the role of creative industries in economic development and urban renewal. This turn to culture, however, has mobilized discourses that privilege particular ideas about what constitutes creativity, notions that are appositely debunked by the arguments of Hallam and Ingold (2007) discussed above. In addition, the assertions of Florida (2005) and other champions of the creative class and the creative city have perpetrated a hierarchal ordering in which places are ranked against one another, a process that includes the identification of 'cool' metropolitan downtown areas and cultural quarters, aestheticizing particular cities such as Barcelona and New York and their bohemian enclaves and cultural quarters. Those places with poor cultural provision, art galleries and upscale cafés are implied to lack those qualities that will attract an aspiring and dynamic creative class who can further develop a reputation for creativity. By implication then, small town, suburban, village and rural forms of making and creativity are sidelined by such assessments and the privileging of the metropolitan and urban. In refuting such value judgements, the small town of Slaithwaite is a setting for practices of making that involves longstanding creative participation, fosters local making practices, contributes to place-making, attracts artistic endeavours and forges links beyond the town.

Moreover, the trumpeting of a creative class redefines the artist as an 'entrepreneur' and risk-taker, thereby sidelining less commercially oriented, alternative and subversive forms of creativity and making. At *Moonraking*, vernacular making practices are privileged. Vernacular creativity possesses the potential to transform everyday space and mundane routine, qualities that cannot be measured in economic terms. Besides providing a source for self-esteem, place identity and community belonging (Burgess, 2006), vernacular creativity imprints class and ethnic identities on place (Markusen, 2006) and transmits skills, inventiveness and humour across localities.

In a current climate of austerity and cuts, small scale cultural events and workshops are difficult to monetize. They are threatened by political neo-liberal and business imperatives to prioritize commercial sustainability, tethering financial support to banal edicts about enterprise, business plans and economic sustainability. However, there is a non-commercial ethos at the heart of many of these modest cultural events. Rather, the creative practices of making they promote are sensuous, creative and convivial, concerned with making objects and places through shared work and performance. These maker cultures celebrate a 'proximate sociality' (Carr and Gibson, 2015: 4) and small town festivals like *Moonraking* are part of what Curran (2010) calls a kind of counter-cultural renaissance in small-scale making. Often emerging in industrial cities and regions where manufacturing has declined, making practices are deployed to reconnect with earlier industrial skills that value local provenance, shared crafting and convivial fun, qualities that have clustered around *Moonraking*.

References

Alves, T. (2007) Art, light and landscape, new agendas for urban development, *European Planning Studies*, 15(9), 1247–1260.

Barker, C. (1999) *Television, Globalization and Cultural Identities*, Buckingham: Open University Press.

Böhme, G. (2010) On beauty, *The Nordic Journal of Aesthetics*, 21(39), 22–33.

Bond, G., Burton, A. and Ellis, K. (2006) *Arking Back, 21 Years of the Slaithwaite Moonraking Festival*, Huddersfield: Reflex.

Burgess, J. (2006) Hearing ordinary voices: Cultural studies, vernacular creativity and digital storytelling, *Continuum, Journal of Media and Cultural Studies*, 20(2), 201–214.

Carr, C. and Gibson, C. (2015) Geographies of making: Rethinking materials and skills for volatile futures, *Progress in Human Geography*, doi:10.1177/0309132515578775

Collins, J. and Jervis, J. (2008) Introduction, in Collins, J. and Jervis, J. (eds) *Uncanny Modernity, Cultural Theories, Modern Anxieties*, London: Palgrave.

Curran, W. (2010) In defense of old industrial spaces: Manufacturing, creativity and innovation in Williamsburg, Brooklyn, *International Journal of Urban and Regional Research*, 34(4), 871–885.

Debord, G. (1994) *The Society of the Spectacle*, New York: Zone Books.

Edensor, T. (ed.) (2010) *Geographies of Rhythm: Nature, Place, Mobilities and Bodies*, Aldershot: Ashgate.

Edensor, T. (2012) Illuminated atmospheres: Anticipating and reproducing the flow of affective experience in Blackpool, *Environment and Planning D, Society and Space*, 30, 1103–1112.

Edensor, T. (2015a) Producing atmospheres at the match: Fan cultures, commercialisation and mood management, *Emotion, Space and Society*, 15, 82–89.

Edensor, T. (2015b) Light design and atmosphere, *Journal of Visual Communication*, 14(3), 331–350.

Edensor, T. (2017) *From Light to Dark: Daylight, Illumination and Gloom*, Minnesota University Press: Minneapolis.

Edensor, T., Leslie, D., Millington, S. and Rantisi, N. (eds) (2009) *Spaces of Vernacular Creativity*, London: Routledge.

Evans, H. (2012) Wonder and the clinical encounter, *Theoretical Medicine and Bioethics*, 33, 123–136.

Fisher, J. and Drobnick, J. (2012) Nightsense, *Public*, 23(45), 35–63.

Florida, R. (2005) *Cities and the Creative Class*, London: Routledge.

Gotham, K. F. (2005) Theorizing urban spectacles: Festivals, tourism and the transformation of urban space, *City*, 9(2), 225–246.

Hall, S. (2001) Encoding/decoding, in Kellner, D. and Durham, M. (eds) *Media and Cultural Studies, Key Works*, Oxford: Blackwell.

Hallam, E. and Ingold, T. (2007) Creativity and cultural improvisation: An introduction, in Hallam, E. and Ingold, T. (eds), *Creativity and Cultural Improvisation*, London: Routledge.

Hawkins, H. (2015) "All it is is light": Projections and volumes, artistic light, and Pipilotti Rist's feminist languages and logics of light, *Senses and Society*, 10(2), 158–178.

Jakob, I. (2012) The eventification of place: Urban development and experience consumption in Berlin and New York City, *European Urban and Regional Studies*, 20(4), 447–459.

Julier, G. (2005) Urban designscapes and the production of aesthetic consent, *Urban Studies*, 42(5/6), 869–887.

Knabb, K. (ed.) (2006) *Situationist International Anthology*, Berkeley, CA: University of California Press.

Markusen, A. (2006) Building the creative economy for Minnesota's artists and communities, *Cura Reporter*, Summer, 16–25.

Micklem, D. (2006) *Street Arts Healthcheck*, London: Arts Council England.

Miller, A. (2016) Creative geographies of ceramic artists: Knowledges and experiences of landscape, practices of art and skill, *Social and Cultural Geography*, URL http://dx.doi.org/10.1080/14649365.2016.1171390.

Papadaki, E. (2015) Curating lights and shadows, or the remapping of the lived experience of space, *Senses and Society*, 10(2), 217–236.

Picard, D. and Robinson, M. (2006) Remaking worlds: Festivals, tourism and change, in Picard, D. and Robinson, M. (eds) *Festivals, Tourism and Social Change, Remaking Worlds*, Clevedon: Channel View.

Pine, J. and Gilmore, J. (1999) *The Experience Economy*, Boston: HBS Press.

Quinn, B. (2010) Arts festivals and urban tourism and cultural policy, *Journal of Policy Research in Tourism, Leisure and Events*, 2(3), 264–279.

Rancière, J. (2009) *Aesthetics and Its Discontents*, Cambridge: Polity Press.

Schulte-Römer, N. (2013) Fair framings: Arts and culture festivals as sites for technical innovation, *Mind and Society*, 12(1), 151–165.

Sloan, S. (2015) Urban subjects, urban luminosity, *The Senses and Society*, 10(2), 200–216.

Stevens, Q. (2007) *The Ludic City: Exploring the Potential of Public Spaces*, London: Routledge.

Stewart, K. (2011) Atmospheric attunements, *Environment and Planning D, Society and Space*, 29(3), 445–453.

Sumartojo, S. (2015) On atmosphere and darkness at Australia's Anzac Day Dawn Service, *Visual Communication*, 14(3), 267–288.

White, M. (1988) Resources for a journey of hope: The work of Welfare State International, *New Theatre Quarterly*, 4(15), 195–208.

Whiteley, G. (2010) New age radicalism and the social imagination: Welfare State International in the seventies, in Forster, L. and Harper, S. (eds) *British Culture and Society in the 1970s: The Lost Decade.* Newcastle upon Tyne: Cambridge Scholars Publishing.

Woodyer, T. (2012) Ludic geographies: not merely child's play, *Geography Compass*, 6(6), 313–326.

4 Modernity, crafts and guilded practices

Locating the historical geographies of 20th century craft organisations

Nicola Thomas

The historical geographies of craft organisations reveal much about the spatial and social formation of the creative economy. This chapter explores the historical geographies of craft societies that grew through the 20th century in the UK, forming to serve the collective needs of craft practitioners and the burgeoning craft movement. By 1970 the spread of craft organisations and societies was such that a Federation of British Craft Societies was formed to enable the craft profession to be represented by a single body. The 1976 members list of the Federation shows that membership organisations were providing support for individuals who were making a living from their craft, either through single-craft specialist organisations (such as The Society of Scribes and Illuminators) or 'mixed discipline' organisations where potters, basket makers, ironmongers, and jewellers might sit alongside makers such as hand weavers, printmakers, book binders, and shoemakers (for example the Society of Designer-Craftsmen). Many organisations that joined the British Federation of Craft Societies also had a geographical similarity: they were regional organisations, with the names that followed defining 'county' administrative geographical boundaries: The Guild of Yorkshire Craftsmen; the Worcestershire Guild of Artist-Craftsmen, Norfolk Contemporary Crafts Committee, and the Cornwall Crafts Association. By 1976 the Federation was supporting forty-three associate and full member organisations, representing over 10,000 craft practitioners (Federation of British Craft Societies, 1976). A common bond between the diverse societies was the value placed on the hand work of craft practitioners who produced original work to their own designs, distinctive from those who were not admitted to these organisations: visual artists and those that involved larger scale manufacturing. The Federation aimed to work as an overarching crafts development organisation to: improve information flows across the sector; enhance retail and selling opportunities; address training needs; support recreational craft pursuits; and to enhance the status of craft. This chapter explores one of the organisations that joined this Federation, the Devon Guild of Craftsmen, to locate the context through which such organisations developed, the way in which they came into being, and to understand the purpose they hoped to serve in supporting craft practitioners.

This chapter offers an analysis of the creative economy that is sensitive to both its histories and its geographies. While much contemporary research around making cultures explores the current practices, there remains a critical need to understand past expressions of creative industries. As Knell and Oakley (2007: p.5) remind us: 'one of the besetting sins of creative industries policy-making is its obsession with the new, its insistence that everything is "changed utterly," and its seeming ignorance, of its own history'. Exploring the founding stories of the Devon Guild of Craftsmen is a trenchant reminder that making and craft practice has a long association with government policy. We learn that while makers working in their studios might feel that they are disconnected from central government policy, they are in fact caught in a nexus where creativity and policy are bound through the support mechanisms that aim to enable thriving creative or economic activities. Indeed, when we look into the development of the Devon Guild of Craftsmen we find the messy entanglements of this nexus: the energies of crafts practitioners, the spirit of the international artistic movement of modernism, the efforts of a government department, and a rural development experiment. The combined efforts resulted in the formation of an organisation that continues to support contemporary crafts practitioners in much the same spirit as the founding members hoped for.

Locating the historical geographies of craft and the creative economy

This chapter stems from observations within a research project addressing the regional governance of the South West of England creative sector undertaken from 2007–2010 by myself, David C. Harvey and Harriet Hawkins. We were exploring how the flows of policy, investment and activities that had been released as a result of the UK Government's adoption of creative industries economic development strategies were affecting lives of those living and working in the region (Harvey et al., 2011 and 2012). We heard much about the proliferation networks, of the ups and downs of networking in rural areas, of networks that did not result in the promised changes, and of the fatigue as policy led finance ran out and networks relied on voluntary labour (Thomas et al., 2013). Some of these conversations were with designer makers who were invested in a range of networks, including some of the very long-established craft guilds in the region. Their conversations interested me as they were using positive registers when they talked about the benefits that their guild membership brought, and the role that the guild played in their lives. They talked about the value of the social relationships that were forged through the guild, and the importance of accessing a high-quality retail market offered through the organisations. Although not everyone was positive about their guild membership it seemed that guilds stood apart from other networking organizations in the region, and they had distinctive qualities that pointed to the need for more engaged research to understand the dynamics that made them into resilient organisations.

Given that little attention has been paid to this group of craft organisations, it is timely to note that the activities that such organisations have traditionally organised are now being aspired to in support of the new broader creative industrial economy, particularly networking, professional development and accessing new markets. Recent sector research by the Crafts Council (2014) demonstrated that in 2012/13 the UK craft sector generated nearly £3.4bn for the UK economy accounting for 0.3% of UK GVA (gross value added). The craft sector is a challenge to define, not least because of the historical legacies from the ways in which the sector has defined itself around distinctive disciplines and artistic traditions (see for example, Greenhalgh, 1997; Adamson, 2007). In recent years the UK Crafts Council has been keen to break open the traditional idea of craft, exploring the broader presence of skilled craft labour in fields from bio engineering to the aeronautical sector, alongside their efforts to better capture the scale of craft micro-businesses to ensure the ongoing support of a small, but thriving economic sector (KPMG/ Crafts Council, 2016). The dynamics of policy and practice within the craft sector reveals the synergies with other creative sector labour practices and also the specific challenges that are more unique to the craft sector, such as the challenges of apprenticeships, the extreme dispersed nature of practitioners, and the difficulties of sustaining an income where the value of hand-crafted labour is undercut by mass production (see Jakob and Thomas, 2015; Luckman, 2015 and Luckman and Thomas, forthcoming).

An interest in regional craft networks links to geographical writing that looks beyond the creative city and creative cluster literatures, to find the 'other geographies' that bind together the creative economy (see Bell and Jayne, 2010; Gibson et al., 2010; Harvey et al., 2012). This literature is often more nuanced around the importance of historical legacies in the contemporary sphere. For example, Warren and Gibson's (2014) work exploring the emergence and development of the surf board industry highlights the attentive connections between materials, hand skills, social and spatial contexts, and the importance of place and connections over time (Warren and Gibson, 2013; see also Carr and Gibson, 2016). This approach of addressing the cultural and historical geographies of the creative economy is highlighted within this chapter, recognising that while makers and markets continually change, the contexts are shaped by the reservoir of past practices, imaginaries and connections that continue to shape the contemporary creative economy.

When one becomes attentive to understanding past expressions of the creative economy, the antecedents of the contemporary creative economy are revealed. Of particular note in relation to the craft guilds are the organisational structures that have now come to more widely govern the sector. The creative field of craft, as traditionally organised and curated by organisations like the guilds, had selection and election committees reviewing new applications for members, with the achievement of membership being a mark of quality for those who successfully gained entry. The field configuring events of exhibitions and shows has become a way in which the standards of craft, inspired by guild practices, are maintained, but also act as events that bring the communities of practice and their audiences together. Just as, for example,

Power and Jansson (2008) describe the form and function of trade shows as field configuring events, the smaller scale 20th century craft exhibitions and the spaces where makers would sell their work are part and parcel of the creative economy (see also Delacour and Leca, 2011 and Comunian, 2017).

Within craft, the dominant narratives that have 'sold' the romance of making have revolved around the studio and the 'hand' of the maker. Revealing the making practices and demonstrating how a lump of clay becomes a bowl has become a critical device through which a maker's work is marketed. Revealing the 'hand-made' nature of the work is particularly important in communicating value, as the laboriousness of making by hand demands a higher price than mass produced goods (Shayles, forthcoming). Curiosity and concern around how crafts skills are communicated, exchanged, and the meanings that become known through embodied practice has also led to a focus on the body in relation to craft practice. This sits alongside curiosity as to how embodied practice, not easily represented through text or media, can be conveyed to others (Marchand, 2010; Ingold, 2013; Paton, 2013; Patchett, 2016). For makers earning a living from their skilled embodied practice, how they sell their work, gain commissions, and continue to work within the current systems of economic exchange, is of fundamental importance. Considerations of the organisations and structures of education, mentoring, workspaces, networks, and display opportunities are equally important to understand as embodied practice, if we are to appreciate a rounded understanding of the geographies of making. Through the lens of the Devon Guild this chapter contributes to knowledge of organisations and interactions that have been developed to support makers to secure their livelihoods, and form a community of practice that promotes their work and provides structures of support.

Framing context: arts and crafts in early 20th century and the Rural Industries Bureau

Craft guilds certainly have an enduring presence in the UK context: they emerged as powerful trade institutions in the early modern period; experienced a revival in association with the Arts and Crafts movement in the late 19th century; and developed in a new form as practitioner-led networks during the 20th century. One of the enduring features of the guild system is the necessary achievement of standard of skill in order to join, recognising that members are professional makers, using their craft skills to secure their livelihood. This requirement of 'standard' in order to join a guild was established within the medieval guild system. Medieval guilds had strict regulations of standards, offered a system for the division of labour in the market, offered a degree of care for the welfare of their members, and sought to enable the intergenerational passing on of skill and standard through the training of apprentices. Although there are debates amongst economic historians on whether these practices fostered innovation or stagnation (see Richardson, 2001; Richardson, 2004), the standard was seen as a mark of trust in the quality of a person's work, and this 'stamp' of approval continues to be a powerful signifier of the guild ideal. Although the guild system was weakened by the time of the industrial revolution in the UK, the 19th century saw a revival of guild ideals through the anti-industrial critiques and the spirit of new-medievalism in the Victorian period. The 19th century socialist Arts and Crafts Movement railed against the demeaning impacts of mass production, and promoted the standards of medieval hand-skilled labour and collective workshop organisation. Such sentiments were found in the writings of John Ruskin and William Morris and through figures such as Charles R. Ashbee, a noted architect and designer of silver and jewellery. For Ashbee, 'The Arts and Crafts movement then, if it means anything, means Standard, whether of work or of life, the protection of Standard, whether in the product or in the producer, and it means that these two things must be taken together' (Ashbee, 1908: p.10). His reflections on standards were written in 1908 shortly after he moved his workshops called the 'Guild of Handicrafts' from London to Chipping Camden in rural Gloucestershire. This enterprise was ultimately to fail, but Ashbee had attempted to run his workshop with a vision of a standard that brought together the value of craft and human labour.

The late 19th century arts and crafts movement saw the development of a series of arts and crafts organisations emerge, imbued with the ethos of collective solidarity and a vision of the importance of beauty, aesthetics, and the ethical treatment of the craftsperson in society. Ashbee's Guild of Handicrafts was a workshop consisting of what he called 'a collective grouping' which saw 'a number of workmen practicing different crafts, carrying out as far as possible their own designs, coming into direct contact with the material, and so organized as to make it possible for the workmen to be wherever necessary in touch with the consumer' (Ashbee, 1908: p.171). Other models were in the vein of a networked organisation, where individual makers would join as a mark of recognition through their membership. The Art Workers Guild in central London, formed in 1884, is an example of an organisation that enabled the gathering of likeminded and skilled practitioners in a city-wide context, to collectivise, to gain support, and inspiration through a programme of education, alongside access to exhibition and display opportunities. It is these models of organisation and the spirit of the aesthetic and socialist principles of the arts and crafts movement that infused the models of craft organisation that evolved in the 20th century. Ashbee reflected that there needed to be formalised governance if standards were to be maintained across the crafts sector. He suggested 'voluntary associations', the 'greater union and organisation of Arts and craftsmen' and direct government legislation that would provide 'general regulation of Industry in the interest of the whole community' as potential routes to secure standards of living and securing the quality of outputs for the crafts sector (Ashbee, 1908: p.91).

These late 19th century organisations were the inspiration for many geographically based guilds which established themselves from the 1930s onwards, under the guidance and encouragement of the Rural Industries

Bureau. The Rural Industries Bureau (RIB) was established in 1921 by the Ministry of Agriculture and was routed through the Development Commission, which the Treasury funded (see Bailey, 1996). Its work was designed to address the challenges experienced by declining rural economies and communities in the post-First World War period. Organised from central London offices, the RIB reached its audiences through printed publications and through a network of staff who were based in county administrative regions across the UK. Their remit was to develop rural industries in their host county by providing technical advice and assistance directly to local business and people looking to develop enterprises. The RIB also had to cope with the diversity of rural industrial activities including farming and agricultural craft trades, and the burgeoning presence of studio based crafts practitioners living in the countryside. The RIB published a quarterly magazine, *Rural Industries: the quarterly magazine for country trades and handicrafts*, written for their desired audience who made a living in the countryside. Within the first decade of this publication you can see the spread of their remit as they published articles to improve the business skills: *Book-keeping is so easy, by J.A.B. Hamilton, Summer 1934*; articles exploring new technologies within established trades: *Oxy-acetylene Welding III Precautions against Expansion and Contraction, by S.L.P. Brewster, Winter 1934*; articles encouraging women and the unemployed to become entrepreneurs and develop home industries based around their craft practice *Thrift Crafts for Women by Alice Armes, Spring 1933; Woodwork for Unemployed Workers III: Cupboards, Sideboards etc., by A Romney Green, Winter 1933*; and articles written by specialists to extend the aesthetic sensibilities of the readership: *Pottery, Decoration in Slip and Metallic Oxides by R.W. Baker, Spring 1935* (Rural Industries Bureau, 1933–1935).

At a county based level, the RIB rural organisers provided practical support direct to their clients. For the craft sector this included business support and the encouragement of county based collective organisations. In counties such as Gloucestershire, the efforts of the RIB staff resulted in the development of the Guild of Gloucestershire Craftsmen in 1933, one of the earliest county based regional organisations, and also provided an outlet for selling makers' work at the RIB offices in Gloucester (Robinson, 1983). The ongoing support of the RIB in Gloucestershire enabled the Guild to establish itself as a broader membership organisation, and establish a regular pattern of temporary exhibitions to enable members to sell directly to the public through a collective event that brought together the range of the members' work (Thomas et al., 2012). Such events can be seen as extensions of the weekly markets that dominated the rural exchange networks (for a review of debates around the historical and economic geographies of periodic markets in the context of the creative economy see Norcliffe and Rendace, 2003) and are the forerunners of the ubiquitous pop up fairs and festivals that now dominate the experience and event economy of the creative sector (Hracs et al., 2013, see also Harris, forthcoming). This periodic, and sustained mode of selling was critical to the success of regional craft guilds as they sought to secure new retail markets in the mid-20th century.

Founding stories: the Devon Guild of Craftsmen

Given that the RIB supported crafts people in nearby counties to establish craft guilds in the 1930s, it is curious that the Devon Guild of Craftsmen did not emerge until 1955. The 'official' history of organisations often starts at the 'first' formal gathering or with the first committee minutes. However, for organisations like the Devon Guild of Craftsmen such 'firsts' are never clear-cut. Although the Devon Guild has a well-preserved set of records that details the activities of the Guild through its history, these archives are fairly silent on the earliest 'founding story' of the Guild. The first substantive record associated with the early years is the first exhibition catalogue of 1956 and this lists the patron, committee and members who displayed their work in the first exhibition. These names give us an entry point to start exploring the connections between people, and the places in which founding members were circulating. The first exhibition catalogue tells us that Leonard Elmhirst was Vice-President, and half the committee members were based out of Dartington Hall, a large rural estate in Devon (The Devon Guild of Craftsmen, 1956). Indeed, many early members of the Devon Guild of Craftsmen had workshop and studio addresses near Dartington Hall. In the absence of paper archives within the Devon Guild of Craftsmen's own records, this research turned to the Dartington Hall archives, now held at the Devon Heritage Centre. These records help us understand the way in which craft and support for rural makers was being addressed in the inter-war period in Devon; and reveals the role of Dartington Hall and the RIB in enabling a rural creative community to collectivise.

Dartington Hall advertised itself as 'an experiment in rural reconstruction' (Rural Industries Bureau, Summer 1935: p.22) through visionary approaches to modern estate management and the intersection of modernist ideas and creative practice. Leonard and Dorothy Elmhirst bought the dilapidated Dartington Estate in the mid-1920s, bringing with them personal finance to invest, and an innovative vision of rural landownership. The letter paper heading of Dartington Hall Ltd. by the mid-1930s notes the business-like nature of the estate, stating its incorporation 'for the purpose of research and rural development' (see, for example a letter from the manager of the Dartington Sawmills to Rex Gardner, 31st July 1935, Rex Gardner Reports, C/RIB/1/C, Devon Heritage Centre). It was a place where two people's vision, and wealth, brought a specific sort of modernity to the British rural countryside: a modernity inspired by international currents of thinking around progressive education, social improvement and aesthetics. It expressed the place of creativity in society, and the connections between arts and economic regeneration. Dartington was seen as a living laboratory and within the first ten years the dilapidated estate had new model farms, architect designed housing estates for workers, signature modernist homes, crafts workshops, dance and theatre companies, a school, garden department, burgeoning art and craft collections, sawmill, textile mill, international conferences and summer schools attracting artists and leading thinkers from all over the world. It was an experiment that had a national and international profile and was linked to an idea of regenerating country house estates in an age of decline (for discussion of the history of Dartington Hall in relation to cultural policy see Upchurch, 2013). The agenda of rural reconstruction was key to the Dartington Hall project. It is therefore of no surprise that the Elmhirsts found a way to link to the national Government rural development agenda delivered through the Rural Industries Bureau.

Given Dartington's profile and its reach one might expect to find a connection between the RIB and Dartington, but the correspondence and records show that it was a complex and close relationship. Unlike other counties where the work of the RIB was organised through the county council committees, the work of the RIB in Devon in the inter-war period was tied into the work of Dartington Hall. In the mid-1930s, we see Rex Gardner, Dartington's Head of Building Works, an architect and wood turner, being employed as the RIB County organizer for Devon. Gardner wrote a letter to the eminent potter, Bernard Leach, on the 2nd February 1934 that details the training he was shortly to undertake as part of the Bureau work, and also points to the uncertainty of his economic position:

> I am required to spend a few weeks with the Rural Industries Bureau touring the Eastern Counties and in their drawing office to learn the tricks of the trade ... [RIB] has no money to spare at present for my salary and I understand that Dr Slater [Dartington Director] managed to persuade L.K.E [Leonard Elmhirst] to stand the strain for 6 months ... I hope I shall not find myself without the R.I.B. job or the studio in six months' time and be landed high and dry.
> (Gardner writing to Bernard Leach, 2nd February1934, Devon Heritage Centre C/RIB/1/C)

It is notable that the RIB was not paying his salary unlike other county coordinators, instead the Dartington Estate covered this. The arrangements about who would pay his salary, and the management of him as a worker continued to be very tortuous throughout the next four years (see Rex Gardner Reports, C/RIB/1/C, Devon Heritage Centre). Gardner was continually uncertain about his job security during this period, particularly as the Elmhirst's were having to retrench and reduce their subsidy to the estate's loss-making activities. Within a few years of Gardner's work, the Managing Director of Dartington was negotiating directly with the RIB for a special dispensation to apply for grants to fund the Bureau's work in Devon. It was noted in Whitehall that is was highly unusual for a private company to be contracted to deliver work of this nature. For Dartington, it was a timely diversification of income, enabling Gardner to continue to work on the estate funded through his government work, whilst raising the profile of Dartington within central government.

Surveying the creative activities within Devon

Rex Gardner was employed to undertake a survey of rural industries in Devon. Although he was supposed to offer marketing and product development advice, much of his time was spent mapping the industries. This was an incredibly slow process and the letters exchanged show that the RIB was anxious for swift results. However Rex Gardner was faced with a large county with a highly dispersed set of makers. When he started his mapping survey he made a special request for a car, petrol budget and two OS maps: one for his wall to flag the places he had visited, and a second to guide him to Devon's many settlements (Gardner, 1934). Gardner's job was to talk to people and gather evidence. He witnessed the rivalries of workshops competing for the declining trade, the isolation felt by workers, he judged people's work on aesthetic and practical grounds, and considered the needs of the industry. Within his regular reports he reflected on what he was finding and his own aesthetic judgements were clear. This is a taste of his experiences of the more commercial potteries he encountered in Devon:

> I have visited several of the rural potteries, which are so distinctive of the county, but in no case have they shown any desire for help. Some are so large that they can hardly be included as rural industries, while all have their own sales organization and are perfectly satisfied with their technical efficiency. Much of their output is artistically bad, but so long as it sells readily it will be difficult to convince the firms concerned that good-class ware might sell even better. Should trade fall off they might be more amenable to suggestions, but meantime they are quite content.
>
> (Gardner, First Report to the RIB, 21st January 1936, Rex Gardner Reports, C/RIB/1/C, Devon Heritage Centre)

We see here the tension between small batch production and artists led craft aesthetics compared to larger scale industrial manufacturing. We also see that at the time of Gardner's visit, such commercial potteries were still viable within the Devon countryside.

Rex Gardner's reports are revealing of the suggestions that were circulating at the time to develop rural industries. He noted the problems of attracting young people to stay in rural crafts such as blacksmithing, and the concern that the skills would be lost, despite the fact that there was still sufficient demand for the work. To overcome this problem he suggested that such industries might cluster into village repair centres to serve a larger area. He was also thoughtful about the need to invest in the marketing work, both in terms of exposure at the large county based shows, but also in niche shops associated with the Dartington Estate (Gardner, First Report to the RIB, 21st January 1936, Rex Gardner Reports, C/RIB/1/C, Devon Heritage Centre).

Gardner also raised issues of the need for cooperative action, supporting rural crafts practitioners at a collective level. Gardner had spent time listening to the issues of makers, and was particularly exercised about what he referred to as 'the evils of credit'. He wanted to encourage a trade association that would offer financial support to makers, and address issues of price-cutting amongst rivals that was undermining income. He saw this as a way of supporting their welfare and sustaining community life. For these issues to be addressed, Gardner turned to his socialist cooperative principles to achieve his aims, and in his report he advocated an approach that: 'entails organizing concerted action on a scale not hitherto attempted; it entails, in fact bringing all the active tradesmen into an organization or a Guild' (Gardner, First Report to the RIB, 21st January 1936, Rex Gardner Reports, C/RIB/1/C, Devon Heritage Centre).

It is in these threads of thought that we start to see the embryonic ideas of a regional craft guild forming. However, Gardner was not suggesting a 'fine craft guild' following the lines of the neighbouring Gloucestershire Guild of Craftsmen and Somerset Guild of Craftsmen (both founded in 1933). Although Rex Gardner was more personally invested in work that was made by who we now might refer to as 'designer-makers', he took his job to support agricultural trades very seriously. The work that Gardner did in Devon gathered momentum in the late 1930s and various one-day events were organised, supported by the Bureau, Dartington estate and Leonard Elmhirst. Gardner organised a Devon Rural Trades Conference, which included his survey results, and showcased the ways in which the RIB could give support to rural crafts' traders. Alongside a film of logging and saw milling and a slide show of Devon trades, the idea of a cooperative organisation was discussed (reported in the Totnes Times, 1st October 1938, Devon Rural Industries, Conference at Dartington Estate, Rex Gardner Reports, C/RIB/ 1/C, Devon Heritage Centre).

Dartington and international creative connections

The onset of the Second World War focused Devon RIB activity on agricultural trades, at the expense of what were referred to as 'luxury products' of craft. Letters between Mr Slater, Director of Dartington, and Mr Marston, the Director of the RIB, demonstrate the wartime conversation around the future of the RIB work was facilitated by Gardner's research:

> Agricultural plough up will suggest that e.g. smiths and wheel wrights will be needed more than ever ... certain of the rural industries which deal largely with luxury products would, I think, have to be closed down ... the change in case of war would be the intensification of that

part of the work which deals with agricultural implements at the expense
of the more artistic sections of your [RIB] work ...

(Letter from Slater to Marston, 30th May 1939, Rex Gardner Reports,
C/RIB/1/C, Devon Heritage Centre)

During the war period Dartington Hall's own arts research department
developed a series of reports and served on national committees that addressed
the ways in which art, design and urban planning might be enrolled in the
post-war reconstruction period (see Upchurch, 2013). Such committees were
part of the discussion from which the Arts Council and the Council for
Industrial Design emerged. The Council for Industrial Design notably hosted
the 'Britain Can Make It' exhibition in 1946 which triumphantly showcased
manufacturing within Britain, including both industrial and artist led craft.
Dartington also had a very close relationship with the new-formed British
Council, particularly working with the Crafts Director Muriel Rose, an
esteemed crafts' collector, curator and gallery owner.

Within Devon the post-war years leading up to the development of the Devon
Guild of Craftsmen in 1955 were associated with a celebration of the work being
produced by artist craftsmen in the county, and cementing the national and
international connections that Dartington brought. There was a continuous
shuttling between the inward-looking perspective on Devon, and the outward
look to its connections beyond the county boundaries. The Elmhirsts' own
international outlook and the creative space of Dartington meant they could
attract prestigious crafts' makers such as Bernard Leach (an internationally
renowned studio potter) to the estate and infuse the ethos of the organisation.
The zeitgeist of the time certainly encouraged this celebration of the local,
amidst the national reconstruction agenda. This was demonstrated by the 1950
'Made in Devon' (Figure 4.1) summer exhibition at Dartington, which celebrated 'beautiful objects' from the past and contemporary makers in Devon that
were valued for rooting the crafts in a particular place (see T Arts Applied 2, B2
Exhibition from 1950, Dartington Hall Records, Devon Heritage Centre). Many
of the contemporary makers whose work was displayed went on to become
founding members of the Devon Guild of Craftsmen. This exhibition was a pilot
for the regional celebrations planned for the Festival of Britain that would take
place in 1951.

This was a burgeoning time within the Dartington Hall estate. The Inter-
national Summer School had been established, and key figures within British
studio craft were connected to the estate's creative vision. These included
Devon Guild founding members including Marianne de Trey who had moved
with her husband T.S. Sam Haile to establish a pottery at nearby Shinner's
Bridge in 1946. Before his untimely death in a road traffic accident (1948),
Haile was working for the RIB as a Pottery Advisor, which further bound
Dartington's connection with the RIB (VADS, n.d.).

Alongside 'Made in Devon' the 1952 'International Conference of Craftsmen' should be seen as a key context out of which the Devon Guild of

DARTINGTON HALL SUMMER EXHIBITION

MADE IN DEVON

An exhibition of beautiful objects, past and present, including Exeter silver, Plymouth porcelain, Church plate and vestments, Honiton lace, woven and printed textiles, earthenware, leatherwork, edgetools, baskets and things made by fishermen.

OPEN DAILY 11–7 JUNE 25 TO AUGUST 27

Admission 1/- (Children 6d) Catalogue 2/-

*A guide lecturer is available by appointment to take parties round the exhibition
Tea (available on weekdays between 4 and 4.30) 1/6*

Enquiries to ARTS DEPARTMENT, DARTINGTON HALL, TOTNES (2272)

Figure 4.1 'Made in Devon' exhibition invitation card, 1950
Source: T Arts Applied 2, B2 Exhibition from 1950, Dartington Hall Records, Devon Heritage Centre, with permission of the Dartington Hall Trust

Craftsmen emerges. The Conference was organised in cooperation with the British Council with Muriel Rose working closely to bring the international delegates to the conference (see D1 International Conference of Craftsmen 1952, Proposal, notes and minutes, Dartington Hall Archives, T Arts Applied 2, Devon Heritage Centre). Opened by Leonard Elmhirst, delegates heard talks from across a range of themes including: Bernard Leach, 'The Contemporary Potter'; Dr S. Yanagi, 'The Japanese Approach to the Crafts'; Michael Cardew, 'The Craftsman's Use of Scientific Development', and Alec Hunter, 'The Craftsman and the Textile Industry' (The Report of the International Conference of Craftsmen in Pottery & Textiles at Dartington Hall, Totnes, Devon, 17–27 July 1952, copy held in uncatalogued papers, Devon Guild of Craftsmen).

The conference press coverage of the time recorded the delegates who had made their way to Britain, including the Catalan potter Jospeh Llorens Astigas (Dartington Hall Archives, T Arts Applied 2, Devon Heritage Centre). Associated with the conference was an exhibition of Pottery and Textiles made in Great Britain between 1920–1952 (ibid.). The Arts Council played an important role funding this exhibition, which went on to tour Edinburgh, London and Birmingham. These activities brought together an international community of crafts' practitioners and gave a visible presence to the ongoing importance of Dartington Hall. Founding members of the Devon Guild of Craftsmen were present at this conference, took part in the exhibitions, and were part of the creative community around the Dartington experiment.

Early years of the Devon Guild of Craftsmen

Returning to the exhibition catalogue of the Devon Guild of Craftsmen in 1956 we are reminded that Leonard Elmhirst is named as the Vice-President, and Dartington Hall based makers including Bernard Forrester, potter, and Edward Baly, furniture maker, were founding members. The Devon Guild papers tell us that committee meetings were frequently held at Dartington Hall (and included Rex Gardner), and the Devon Guild regularly held weekend gatherings hosted at the Hall to bring the dispersed members together. The discussions around the role of cooperative working that Gardner raised in the late 1930s were brought into the ethos of the Devon Guild from the start. We can also see that the arts and crafts movement ethos is firmly present in the founding of the Devon Guild of Craftsmen: fine craft skills, with the maintenance of quality, standards and aesthetics regulated through strict membership criteria and a bringing together of like-minded crafts practitioners to sell their work through collective endeavours. The connections between the arts and craft movement were brought into the Guild through another founding member, Judith Hughes, furniture maker, whose mother trained under Charles Ashbee and George Hart in the Guild of Handicrafts in Chipping Campden, Gloucestershire.

The Devon Guild of Craftsmen's founding principles, as set out in the 1959 Summer exhibition catalogue, point to the desire of the Guild for the membership to support makers' livelihoods by creating a retail opportunity, but also in pushing the boundaries of craft: 'At the time of founding the Guild, it was agreed that the best way of providing this encouragement [to crafts practitioners] was to create a market, by means of Exhibitions, for members' current work; and in particular for work of a more experimental kind not readily acceptable by the normal outlet open to craftsmen [sic.]' (The Devon Guild of Craftsmen, 1959).

The 'Aims of the Guild' go on to state that 'Membership of the Guild is open to all practising craftsmen ... the applicant must satisfy the Council of the Guild as to their technical and creative ability' (ibid.). The concerns with quality craftsmanship and 'standard' defined by the Guild in its founding decade continue to be a key element of the ethos of the organisation (with the outmoded gendered expression of the 'craftsman' [sic.] causing tensions for some contemporary members). This distinction round fine craft had always been a tension, as evidenced in the commentaries of Rex Gardner as he went around the rural crafts workshops. He wanted to raise the standards of craft production, however, as an artist's craftsman himself he had an affinity to those who were engaged with the development of the aesthetic sensibilities of their work. He did, however, understand the need to support the broader agricultural and rural

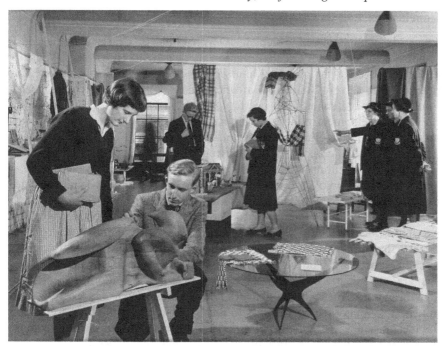

Figure 4.2 The Devon Guild of Craftsmen Summer Exhibition, 1958
Source: Uncatalogued papers, with permission of The Devon Guild of Craftsmen

crafts, with a more open, and cooperative approach. The emergence of the Devon Guild of Craftsmen saw the cementing of the support of an exclusive arts and crafts tradition that valued fine craft skills, with the maintenance of quality, standards and aesthetics, regulated through strict membership criteria.

The early activities of Devon Guild set a strong precedent. The annual summer exhibitions were selling opportunities: there were craft demonstrations and curated exhibitions that customers could imagine in their own home, and they could then go on to purchase or commission items to take home. The minutes of the meetings record the labour that went into making these avenues to market: identifying exhibition spaces, organising access, insurance, marketing, designing the exhibition, organising delivery of items to be displayed, ensuring the items were of the right quality for display, stewarding, and financial accounting of sales. It has to be remembered that in the context of the 1950s the avenues of selling direct to your local audience were limited. Guild exhibitions were part of the seasonal calendar and the periodic selling exhibitions continued until 1986, when members of the Devon Guild invested their own money to purchase the Riverside Mill in the village of Bovey Tracey and set up their permanent gallery. In one swift shift, the guild moved from being a practitioner-led organisation to one which required staff to manage the building and raise money to secure the building and the future of the guild.

Conclusions

The early context out of which the Devon Guild of Craftsmen emerged is important for the history of the Guild in its own right. However, this history also reminds us of the on-going intersection of economic policy and creative practice. The Devon Guild of Craftsmen emerged from the spirit of international modernism encouraged on the Dartington Hall estate, combined with the rural development policies of the Rural Industries Bureau. The entanglements of local and global currents of thinking around craft aesthetics and function show the complex place that regions like Devon have in the development of 20th century craft and design. They also allow us to consider how a very specific vision of craft was enrolled, enabling highly skilled and dedicated makers to sell and exhibit their work, within a cooperative guild.

The Devon Guild of Craftsmen has now been established for over 60 years, but it is still tied to the ethos of the early years. It continues to define its membership through the quality of a person's skilled practice which fits within disciplines of craft. The boundaries of craft are stretched as craft practice changes, with digital tools and technical materials being adopted by makers, and the guild navigates these changes within the selection and election committee. It continues the opportunity to provide members with access to support for their business, particularly for emerging makers who are starting to make their living through their craft practice. The guild continues to exhibit contemporary craft in its gallery spaces, as well as providing sales opportunities within Riverside Mill as well as online. The guild has retained its international perspectives, with international exhibitors and exchanges happening within its membership. In addition, the guild continues to be bound into a creative policy environment funded by the UK Government and independent trusts and foundations. Indeed, as an Arts Council National Portfolio Organisation it receives regular income which places it very centrally as one of an elite set of cultural organisations in the UK that receives financial support underpinned by the UK Government. The ties with Dartington Hall continue to be present with guild members teaching and working in studios on the estate.

The geographies of making that are entangled in this story are ones that reveal the webs of connections within a creative community of practice. Thinking about the quality of these connections reveals that it has been individuals working together that made the Devon Guild come into being. Although Dartington Hall and the Rural Industries Bureau were bound together, it was actually the strength and persistence of the personal relationships between individuals in the organisations that forged the relationship, and delivered the national policy trends at a regional level. The richness of the creative conversations and practice that emanated from Dartington Hall attracted the attention and enthusiasm of practitioners and funders who saw it as a place where something would happen. The vision of Dorothy and Leonard Elmhirst, and the people who were employed to bring this vision into reality created opportunities for the conversations to take place that resulted in a guild forming. As this chapter has indicated, these conversations did not happen by accident, they evolved from over twenty years of activities that provided the foundations for the Guild to emerge and develop into a secure organisation that was very sure of its purpose, and with a membership that was committed to securing its future.

Acknowledgements

I would like to acknowledge and thank the Arts and Humanities Research Council for funding the projects that have underpinned this research (AH/E/008887/1 and AH/I001778/1). I would also like to thank the staff and membership of the Devon Guild of Craftsmen and Gloucestershire Guild of Craftsmen for their support of the research. My thanks to Doreen Jakob, David C. Harvey, Harriet Hawkins, Chris Gibson and audience participants at the International Conference of Historical Geographers.

References

Adamson, G. (2007) *Thinking through Craft*. Oxford: Berg.
AshbeeC. R. (1908) *The Guild of Handicraft, its Deed of Trust and Rules for the Guidance of its Guildsmen, together with a Note on its Work*. Broad Campden: Guild of Handicraft.
Bailey, A. (1996) Progress and preservation: The role of rural industries in the making of the modern image of countryside. *Journal of Design History*, 9(1), 35–53.
Banks, M. (2010) Craft labour and creative industries. *International Journal of Cultural Policy*, 16(3), 305–321.
Bell, D. and Jayne, M. (2010) The creative countryside? Policy and practice in the UK rural cultural economy. *Journal of Rural Studies*, 26(3), 209–218.
Carr, C. and Gibson, C. (2016) Geographies of making: Rethinking materials and skills for volatile futures. *Progress in Human Geography*, 40(3), 297–315.
Comunian, R. (2017) Temporary clusters and communities of practice in the creative economy: Festivals as temporary knowledge networks. *Space and Culture*, 20(3), 329–343. Available from https://doi.org/10.1177/1206331216660318 [Accessed 10th September 2017].
Crafts Council (2014) *Measuring the Craft Economy: Defining and Measuring Craft: Report 3*. Available from http://www.craftscouncil.org.uk/content/files/Measuring_the_craft_economy-v4.pdf [Accessed 12th September 2017].
Delacour, H. and Leca, B. (2011) A Salon's life. Field configuring event, power and contestation in a creative field. In Moeran, B. and Strandgaard Pedersen, J. (eds), *Negotiating Values in the Creative Industries: Fairs, Festivals and Other Competitive Events*. Cambridge: Cambridge University Press.
Federation of British Craft Societies (1976) *Annual Members List*. Available from Museum of English Rural Life, MERL Library Pamphlets, 5610, Box 1/01.
Federation of British Craft Societies (1976) *Federation and Membership Information*. Available from the Museum of English Rural Life Library Pamphlet 5610 Box/01.

Gardner, R. (1934) Memo to Dr Slater 2/7/1934, *Survey of Devon Craftsmen*. Available from Dartington Hall archives, Rex Gardner Reports, C/RIB/1/C, Devon Heritage Centre.

Gibson, C., Luckman, S. and Willoughby-Smith, J. (2010) Creativity without borders? Rethinking remoteness and proximity. *Australian Geographer*, 41(1), 25–38.

KPMG/Crafts Council (2016) *Innovation through Craft: Opportunities for Growth*. Available from http://www.craftscouncil.org.uk/content/files/KPMG_CC_innovation_report_full.pdf [Accessed 10th September 2017].

Greenhalgh, P. (1997) The history of craft. In Dormer, P. (ed.) *The Culture of Craft: Status and Future*. Manchester: Manchester University Press.

Harris, E. (forthcoming) Crafted places/places for craft: Pop-up craft spatialities and the politics of inclusion. In Luckman, S. and Thomas, N. J. (eds), *Craft Economies*. London: Bloomsbury.

Harvey, D. C., Hawkins, H. and ThomasN. J. (2011) Regional imaginaries of governance agencies: Practising the region of South West Britain. *Environment and Planning A*, 43(2), 470–486.

Harvey, D. C., Hawkins, H. and Thomas, N. J. (2012) Thinking creative clusters beyond the city: People, places and networks. *Geoforum*, 43(3), 529–539.

Hracs, B., Jakob, D. and Hauge, A. (2013) Standing out in the crowd: the rise of exclusivity-based strategies to compete in the contemporary marketplace for music and fashion. *Environment and Planning A*, 45, 1116–1144.

Ingold, T. (2013) *Making: Anthropology, Archeology, Art and Architecture*. Abingdon: Routledge.

Jakob, D. and ThomasN. J. (2015) Firing up craft capital: The renaissance of craft and craft policy in the United Kingdom. *International Journal of Cultural Policy*, 23(4), 495–511.

Knell, J. and Oakley, K. (2007) *London's Creative Economy: An Accidental Success?* Provocation Series. London: The Work Foundation.

Luckman, S. (2015) *Craft and the Creative Economy*. Basingstoke: Palgrave Macmillan.

Luckman, S. and Thomas, N. J. (eds) (forthcoming) *Craft Economies*. London: Bloomsbury.

DartingtonHall (1950) *Made in Devon*. Available from Dartington Hall Records, Devon Heritage Centre.

Marchand, T. H. J. (ed.) (2010) Making knowledge: Explorations of the indissoluble relation between mind, body and environment. *Journal of the Royal Anthropological Institute*, 16 (May). Also available at doi:10.1002/9781444391473.index [Accessed 10th September 2017].

Norcliffe, G. and Rendace, O. (2003) New geographies of comic book production in North America: The new artisan, distancing, and the periodic social economy. *Economic Geography*, 79(3), 241–263.

Patchett, M. (2016) Taxidermy workshops: Differently figuring the working of bodies and bodies at work in the past. *Transactions of the Institute of British Geographers*, 42(3), 390–404.

Paton, D.A. (2013) The quarry as sculpture: The place of making. *Environment and Planning A*, 45(5), 1070–1086.

Power, D. and Jansson, J. (2008) Cyclical clusters in global circuits: Overlapping spaces in furniture trade fairs. *Economic Geography*, 84(4), 423–448.

Richardson, G. (2001) A tale of two theories: Monopolies and craft guilds in medieval England and modern imagination. *Journal of the History of Economic Thought*, 23(2), 217–242.

Richardson, G. (2004) Guilds, laws, and markets for manufactured merchandise in late-medieval England. *Explorations in Economic History*, 41, 1–25.

Robinson, S. (1983) *A Fertile Field: An Outline History of the Guild of Gloucestershire Craftsmen and the Crafts in Gloucestershire.* Gloucester: The Guild of Gloucestershire Craftsmen.

Rural Industries Bureau (1933–1935) *Rural Industries: The Quarterly Magazine for Country Trade and Handicrafts.* Available from the Museum of English Rural Life, Museum of English Rural Life. Library Periodical Open Access – PER.

Rural Industries Bureau (Summer 1935) *Advert for Dartington Hall Published in the Rural Industries: The Quarterly Magazine for Country Trade and Handicrafts.* Available from the Museum of English Rural Life, Museum of English Rural Life. Library Periodical Open Acess – PER.

Shayles, E. (forthcoming) The ghost potter: Vital forms and spectral marks of skilled craftsmen in contemporary tableware. In Luckman, S. and Thomas, N. J. (eds), *Craft Economies.* London: Bloomsbury.

The Devon Guild of Craftsmen (1956) *First Exhibition of Fine Crafts.* Available from the Dartington Hall Records, LKE Devon 4, J Devon Guild of Craftsmen, Devon Heritage Centre.

The Devon Guild of Craftsmen (1959) *Annual Exhibition at Birdwood House, Totnes.* Uncatalogued exhibition brochure available from The Devon Guild of Craftsmen.

Thomas, N. J., Harvey, D. C. and Hawkins, H. (2012) Crafting the region: Creative industries and practices of regional space. *Regional Studies*, 47(1), 75–88.

Upchurch, A. R. (2013) 'Missing' from policy history: The Dartington Hall Arts Enquiry, 1941–1947, *International Journal of Cultural Policy*, 19(5), 610–622. Available from doi:10.1080/10286632.2012.724065 [Accessed 10th September 2017].

VADS (n.d.) *T. S. (Sam) Hailes (1908–1948).* Weblog. Available from https://vads.ac.uk/learning/learndex.php?theme_id=cscu1&theme_record_id=cscu1haile&mtri=cscu1ceram [Accessed 10th September 2017].

Warren, A. and Gibson, C. (2013) Crafting regional production: Emergence, crisis and consolidation in the Gold Coast surfboard industry. *Australian Geographer*, 44(4), 365–381.

Warren, A. and Gibson, C. (2014) *Surfing Places, Surfboard Makers: Craft, Creativity and Cultural Heritage in Hawai'i.* Hawai'i, United States: University of Hawai'i Press.

5 Unpicking the material politics of sewing for development

Sex, religion and women's rights

Zoe Collins

Introduction

The sewing machine and the needle and thread have become so ubiquitous in development interventions that we perceive them as benign, apolitical objects and do not challenge their deployment. However, as feminist scholars argue, it is those things that appear mundane that most require unpacking (see Dowler and Sharp, 2001; Beaudry, 2006). On GlobalGiving, the non-profit crowdfunding platform, there are currently 57 sewing initiatives seeking funding. Each of these targets women and girls as the primary, if not sole, recipients of training and/or equipment. They also harness the same rhetoric, positioning sewing as a means of generating income and, in turn, enabling empowerment, self-reliance and prosperity. Within development in general, it is well understood that this formula is overly simplistic; as academics such as Pearson (2007) and Chant (2014) have discussed, paid work, whether performed within or outside of the home, does not necessarily improve the economic or socio-cultural position of women. However, such an approach is also problematic because it utilises sewing apparatus without considering the distinct material politics of these items.

It is increasingly understood that rather than viewing material artefacts as simply "the passive and stable foundations on which politics takes place", we must recognise how the "unpredictable and lively behaviour of such objects... [are] integral to the conduct of politics" (Barry, 2013: pp.1–2). Until Rozsika Parker's (1984) *The Subversive Stitch: Embroidery and the Making of the Feminine*, needlework was trivialised and academically neglected due to its uncritical association with women's domestic work (Goggin, 2009). By examining the historical processes within which embroidery became imbued with docility, meekness and servitude, Parker (2010) reveals how the needle has been harnessed to inculcate feminine ideals as part of the dissemination of patriarchal ideology. Thus, although women's relationship with embroidery has been naturalised, it is by no means innate. The understanding that needlework both constructs, and is constructed by, gendered norms, has laid the foundations for Beaudry (2006), Goggin and Tobin (2009) and other vital research into the material politics of sewing. These studies explore the cultural significance of findings (Beaudry, 2006) and the fibre arts that they are used to produce (Goggin and Tobin, 2009), whilst collectively demonstrating that "material culture is not just something people create but an integral component of our personalities and our social lives, deeply implicated in how we construct social relationships" (Beaudry, 2006: p.7). Yet, despite such progress, research has largely been historically situated and disproportionately focussed on Europe and North America. This chapter seeks to reorient the geographies of this work to begin to unpick sewing as a development intervention in the Global South. As Smirl (2011) states, the outcomes of development assistance are influenced by the objects through which programmes are delivered. With this in mind, I will develop three key case studies: Cambodian sex workers, faith-based organisations, and the Adithi *sujuni* project, a women's rights-based initiative. Through these examples, I explore the materials of sewing-based development initiatives as caught within global circuits of trade and aid, as being mobilised within problematic ideologies about sewing as 'moral' work, and as requiring conceptualisation as creative practice. In doing so I will demonstrate that not only do sewing apparatus and practices possess a distinct politics but that, by failing to recognise this, we are allowing the proliferation of actively harmful initiatives.

Sewing and sex work

Sewing machines have a complex place in the history of women's emancipation, the evolution of capital and the processes of globalisation (Beaudry, 2006; Domosh, 2006; Siddiqi, 2009; Prentice, 2017). More recently, sewing has come to have a controversial role in development practices, used as a means of 'saving' those engaged in what is perceived as 'indecent' sex work. Indeed, so prevalent is this practice that The Asian Pacific Network of Sex Workers (APNSW) have adopted the logo of 'no sewing machines' (Figure 5.1) in their fight for improved rights and legal protection. In this first section, I seek to explore the case study of Cambodia, where anti-trafficking NGOs, such as AFESIP[1] and Agape International Missions[2], provide sewing lessons as a central component of their sex worker rehabilitation programmes. Such initiatives are problematic for a number of reasons, not least because programme participants are typically recruited through 'raids and rescues'. These entail police officers raiding brothels and forcibly removing women in order to 'save' them from pimps and overlords (Overs, 2009). 'Rescued' women are then taken to the local anti-trafficking department where those who cannot bribe their way to freedom are faced with just two options: agree to re-training for a new career – typically in the garment industry – or remain in custody indefinitely, vulnerable to rape and other police brutality (Alvi, 2014; Overs, 2009). Under the 2008 Cambodian Law on Suppression of Human Trafficking and Sexual Exploitation, all aspects of commercial sex are illegal, irrespective of whether sex workers are consenting adults (Overs, 2009). This creates a rigid dichotomy whereby those who sell sex are seen as *either* trafficked victims *or* criminals

Figure 5.1 'Don't talk to us about sewing machines, talk to us about workers' rights'
https://apnsw.info/

(Hoefinger in Winn, 2016). As a result, the rejection of rehabilitation is viewed as an omission of guilt, whilst its acceptance, by default, is interpreted as confirmation of the victimhood that is assigned to these women. This enables the Cambodian government to claim that they have saved thousands of trafficked women, despite the fact that sex workers are never asked if they required rescuing in the first place (Alvi, 2014). Likewise, it is increasingly recognised that fundraisers and media sensationalists have significantly exaggerated the prevalence of sex trafficking in Cambodia, with published figures often 75 times greater than the number of documented cases (Moore, 2016). Although still a grave and widespread issue, this means that many of the 'rescued' women are not victims of sex trafficking, but rather the assumed confluence of sex work *with* trafficking.

The suppression of sex workers' agency only intensifies as they enter rehabilitation programmes. Despite promises that re-training will enable them to transform their lives, the NGOs that sex workers are transferred to are described as prison-like, with the women locked in and receiving little, if any, money for their sewing work (Pholly in Alvi, 2014; Moore, 2016). Unable to financially support their families, this forces a number of individuals to flee before their training is complete (Pholly in Alvi, 2014). Yet, for those that do make it through, the resulting jobs in the garment industry offer little means of salvation. Chronically underpaid and dangerously overworked, the minimum wage for a Cambodian garment worker is just US$153 a month (Chan Thul, 2016). At 32% of the living wage for a single-earner household (Labour Behind the Label and Community Legal Education Centre, 2013; adjusted to inflation by Fair Labour Association, 2016), this amount – if even received – is barely sufficient to cover basic sustenance and a mattress on the factory floor (Alvi, 2014; Larsson, 2016; Microfinance Opportunities, 2017). Consequently, many women return to sex work, either full time or to supplement their factory wages. At the same time, for those who have been trafficked, the stressful and unsafe environment of the garment industry is hardly conducive to recovery. Without adequate long-term support, including counselling and health care provision, these victims become extremely vulnerable to re-trafficking (US Department of State, 2014).

The inability of garment production to provide a sustainable alternative livelihood raises a fundamental question: who is truly benefitting from the advocation of sewing work? Re-training programmes are derived from "the twin assumptions that no woman would willingly sell sex and that sex workers lack education and skills for 'decent' work" (Overs, 2009: p.23). However, women often state that they could sew prior to being 'rescued' and had entered the sex trade because it was more economically viable than manufacturing textiles (Hunter in Overs, 2009). As such, many are being forced back into the very industry they were running from in the first place. The failure of NGOs to recognise this further demonstrates the neglect of sex workers' voices. However, it is also symptomatic of the garment industry's prominence within Cambodia (Moore, 2016). It is by no means coincidental that the sewing machine has been selected as the main tool for sex worker rehabilitation. Many of the anti-trafficking NGOs work in direct partnership with local officials and some, including the prolific International Justice Movement, also receive government funding (Moore, 2016). As the garment industry makes up 80% of Cambodia's exports (Better Factories Cambodia, 2017) and is dependent on a continuous supply of cheap, largely female, labour, there is a clear state interest in channelling women into this sector. A number of the same NGOs also have ties with the very apparel companies that profit from the country's garment workers. For example, one of the board members of Agape International Missions is Ken Peterson, the CEO of Apricot Lane, a US fashion chain whose garment production is sub-contracted to Cambodian factories (Moore, 2016). This demonstrates the circular relationship between the garment industry and the sex trade, whereby low wages in the former push women into the latter, only for them to be forcibly re-incorporated through rehabilitation initiatives.

At the same time, the sewing machine is embroiled in a wider narrative of American imperialism. Under the 2003 US President's Emergency Plan for AIDS Relief (PEPFAR), governments and organisations that receive US state funding must implement, and actively pursue, "a policy explicitly opposing prostitution and sex trafficking" (Overs, 2009: p.29). As a country heavily dependent on foreign aid, and with the US a principal donor (BEAPA, 2016), this legislation helps to explain Cambodia's blanket illegalisation of commercial sex work, as well as the proliferation of 'raid and rescue' initiatives. It is important to recognise that PEPFAR is part of a wider American crusade to eradicate sex work and save overseas women. This is based on the perception that all sex work is inherently exploitative (US Press Secretary, 2003), an ideology that is then exported to other nations through the mandates of foreign assistance. However, US anti-trafficking policy is also rooted in traditional protestant ideals of "sexually pure and pious womanhood" (Zimmerman, 2013: p.116). The key political role played by conservative evangelical Protestants, particularly within the Republican Party, has led to pronounced gender and sexual conservatism in US policy (Bernstein and Jakobsen, 2010). Increasingly, this conservatism has merged with neoliberal imperatives to produce pro-business approaches to tackling sex work (Bernstein and Jakobsen, 2010). As can be seen in Cambodia, such initiatives are punitive towards women and undermine efforts for gender equality. Indeed, many anti-trafficking activists no longer position the sex industry within the wider dynamics of globalisation, migration and gendered labour. Instead, prostitution is perceived as a humanitarian issue, one that capitalist forces can help to overcome (Bernstein and Jakobsen, 2010).

Yet, as Hoefinger (in Winn, 2016) argues, these are the very forces that trap women in sex work in the first place. There is a social expectation, particularly if they are the firstborn child, that Cambodian women will be 'dutiful daughters' and support their extended families (Ditmore, 2014). However, for those with little education and from poor rural areas, options for work are largely limited to agricultural labour, garment production and sex work (Hoefinger in Winn, 2016). As all of these occupations are low paid and exploitative, women often end up engaged in the latter simply because it is more profitable. The problem is that rehabilitation programmes are based on the assumption that sex workers are somehow lacking and must be 'restored', whether that is in skills, morals or spiritual value. Not only does this reinforce prejudice and self-stigma (Israel et al., 2008), but it fails to recognise, let alone address, the fact that it is Cambodian society that is deficient, both in its stringent gender norms and limited work opportunities for women. Positioning the sewing machine as the 'fallen' woman's redemption then enacts the additional harm of legitimising the existing labour conditions in the garment industry. Clearly, the 'no sewing machines' logo of the APNSW could not be more appropriate.

Sewing 'saves souls' – sewing machines for Jesus

Northern India is one of the least evangelized sections of the country, where missionaries face opposition from Hindus, Muslims, and Sikhs. Praise God for the sewing centers and other community-based programs that are breaking down the walls of hostility and making the gospel message accessible to all.[3]

The Evangelist Christian Aid Mission takes great pride in the fact that 60% of the women who attend their Indian sewing centres have 'received Christ' into their lives. Indeed, their website tells a series of conversion stories based around sewing 'garments for God's Glory', in which vocational sewing classes double as sites to 'share the gospel' with young women who are primarily being taught sewing skills as a means to support themselves and their families. Yet, their website also – seemingly without concern – tells stories of the disruptive effects of such conversion practices, detailing, for instance, how one student was beaten by her Hindu parents after she attended a Christian church under the direction of her sewing teacher. As the website narrates, however, the family and community repented when her mother was taken ill after denying her daughter's attendance at church. Her condition only improved when they 'opened their hearts to the gospel'.[4]

It is perhaps not surprising that many of the charities that provide sewing lessons for development purposes are Christian-orientated. Sewing has a long history of being taught by missionaries to instil Christian norms of femininity in 'foreign' women. Requiring "clean hands, self-sacrifice, and self-discipline", the needle and thread were seen as the ideal apparatus for the inculcation of cleanliness and godliness (Beaudry, 2006: p.113). Likewise, the repetitive nature of sewing was harnessed for religious instruction, with women and girls throughout the colonies tasked with embroidering Biblical scenes and verses (Beaudry, 2006). Sewing skills were also vital for the production of garments that adhered to European standards of modesty and could be worn to church (Thomas, 2002). Moreover, Higgs (2016) discusses how, throughout Africa, vocational training was used to reinforce a gendered division of labour. The roles that were deemed appropriate for women were determined by British and European patriarchal expectations, as well as how missionaries and colonisers perceived African cultures and capabilities (Higgs, 2016). Sewing skills were seen as a necessity for European women and girls because it kept their hands occupied, preventing the idleness that could incite inappropriate behaviour (Beaudry, 2006). This was seen as particularly relevant for African women because they were believed to be extremely impressionable and liable to immorality (Kanogo, 2005). Such characteristics were, in turn, used to justify the domestic and maternal focus of their vocational education (Higgs, 2016). However, as Robert (2008) explains, vocational training was as much about preparing women for motherhood, as it was about ensuring a skilled domestic labour force that could meet the demands of colonial authorities.

Missionaries also used sewing as a way to bring women together and expose them to Christian teachings. The use of the sewing class as a platform for evangelism was particularly important in communities resisting the dissemination of the gospel (Mokosso, 2007). It is also a practice that continues today. The MTW, for example, the global missions agency of the Presbyterian Church in America, has a project called 'Sewing for Jesus'. Of one Thailand-based sewing collective, the organisation proudly boasts: "the [ultimate] purpose of [Napada] isn't to make bags... [but] to share the Gospel with each of our ladies and our customers."[5] The Indian sewing schools of the Worldwide Proclamation, the overseas branch of Australian Ministry Open Air Campaigners, offer another example. Run by Christian sewing teachers, each class begins with Bible study and devotions as the women "learn to pray to the living God, sing Christian Choruses, and hear Bible teachings" (Worldwide Proclamation, 2015). The organisation states that no student is required to convert to Christianity in order to graduate. However, conversion is clearly encouraged given the celebration of students who have become Christian. Likewise, on the Worldwide Proclamation (2015) website, one of the sewing teachers asks for prayer to support students who "are close" to accepting Christ as their Saviour. While the website foregrounds the need for women to learn a "special skill" to "break out of poverty", the organisation also praises the Kanataka (SW India) sewing school, where, of the 22 students, "6 accepted Jesus as Saviour" and "4 students have taken baptism". The stories told are of women without hope, who will "never forget... that Christians made it possible for them to learn to sew... that Christ made it possible for them to have salvation through faith in Him."[6] Of one young woman, a teacher observed, "she is a good learner and listener in the class. Only through this opportunity she feels that she has got a bright future, and Lord God has lighted [sic] a lamp in her life. She is very much thankful also to you for this institute. She accepted Jesus as her Saviour on 28/7/2016. She believes deeply, and has a new life. She is very good, and listens daily to Bible lessons."[7]

It cannot be denied that religion has, and will continue to have, an important role in international development, not least because it is a crucial source of support, comfort and motivation for those experiencing suffering and injustice (see ter Haar, 2011; Clarke, 2013; Tomalin, 2013). However, the predominately Hindu and Muslim women who come to these classes do so to learn sewing, not Christian instruction. By requiring them to engage with the latter in order to access the former, Worldwide Proclamation are exploiting their power as a provider of vocational training in impoverished areas. Furthermore, the organisation appears to have given no consideration to how conversions to Christianity will be received. Many of the accounts discuss how the students come from strong Hindu families and have broken with tradition to convert. Such opposition, as shown by the example of Christian Aid Mission, clearly leaves the women vulnerable to violence, abuse and social exclusion. Whilst more work needs to be done on these examples in particular, it is evident from academic scholarship on missionary work in India that there is an established history of Christian conversions dividing families, neighbourhoods and entire communities, leading to, amongst other consequences, ostracism, disinheritance and the removal of children and/or spouses (Mallampalli, 2004).

Sewing as a 'magic bullet'

Worldwide Proclamation (2017) state that women who are certified in sewing "will" be able to find employment or start their own businesses, bringing both themselves and their families out of poverty. Yet, the organisation provides no evidence of any follow-up research to prove this, and the only information that is given about programme graduates is in regard to their continued commitment to Christ. One of the main reasons for this is that Worldwide Proclamation, by their own admission, view the conversion to Christianity as "more important" than the gaining of sewing skills (Worldwide Proclamation, 2017). However, I also believe that it is symptomatic of a wider, incredibly problematic discourse, in which the ability to sew is being viewed as some sort of 'magic bullet' for the alleviation of poverty. Worldwide Proclamation is one of a number of organisations that are failing to follow-up on their sewing initiatives because they do not feel there is a need. This is because they have absolute confidence that the skills they teach will deliver prosperity. In other words, the notion of sewing oneself out of poverty is a forgone conclusion. Yet, the ability to sew is not a solution in and of itself; for individuals to improve their situation, these skills have to be applied, and there are a number of barriers to doing this successfully. For example, in order to harness sewing capabilities for employment, work opportunities must be available. However, Worldwide Proclamation do not provide jobs for the women that they train, or give any indication of market research to show that there are external employers in demand of these skills. At the same time, even if such work is available, skills alone do not make it accessible. Indeed, due to structural discrimination, individuals may not be of the 'appropriate' sex, caste or religion for the employer to hire. Similarly, childcare and domestic responsibilities can prevent women from having sufficient time to commit to employment. Due to the pervasiveness of India's patriarchal customs, they may also need to seek permission from male household heads, particularly if the work is being performed outside of the home (Rehman, 2010).

On the other hand, if programme graduates are to set up their own sewing businesses, it is essential that they have access to core materials and equipment, such as fabric, sewing machines, cotton reels, needles and tape measures. Yet, Worldwide Proclamation does not provide these items, nor offer any form of credit lending for their purchase. Likewise, whilst Christian Aid Missions (2014) do provide sewing machines to "some" of their students, sustainable businesses require a constant and affordable supply of materials, as well as access to a machine-fixing service, complete with spare parts. There is also a fundamental requirement for a consistent level of customer demand, some-thing which is shaped by the women's social capital and the degree to which their tailoring services are competitive. The problem is that both Worldwide Proclamation and Christian Aid Missions are equipping a large number of women with identical skillsets. This can quickly saturate the market, limiting the profitability, and overall viability, of individual enterprises. Moreover, part of the reason why the ability to sew is expected to alleviate poverty is because the poor are perceived as inherently entrepreneurial. This is grounded in the neoliberal notion of the "entrepreneur of the self" (Rose, 1999: p.142) – "the self-realizing, responsible and calculating market participant who is empowered to take control of his or her own destiny" (Schwittay, 2011: p. S72). Nonetheless, as Karnani (2009: p.81) argues, "the vast majority of the poor lack the skills, vision, creativity and drive of an entrepreneur". Indeed, a successful sewing enterprise requires creative talent and business acumen, however neither organisation provides any form of business training. There is also a substantial difference between following specific class instruction to produce a simple, standardised garment, and turning one's own design, or that of a client's, into an individualised, fitted product. Although marginalised and devalued through its association with femininity and domesticity (Parker, 2010), it must not be forgotten that sewing is a skill and there will always be some individuals who are more accomplished than others.

Sewing as Craftivism – quilting livelihoods, stitching politics

In each of the case studies discussed, sewing has been regarded as purely functional, a means of achieving particular outcomes, delivering specific ideologies, or instilling desired characteristics. There has been no concept of sewing as a creative practice, let alone any understanding of how creativity can be utilised to achieve development objectives. This is largely because sewing has not been taught to stimulate agency, with women too often positioned as passive recipients of sewing instruction. Adithi's *sujuni* project represents an entirely different approach. Based in Patna, the capital of India's Bihar state, Adithi is an NGO that seeks to support women's socio-economic development.[8] Through Mahila Vikas Samyog Samiti (MVSS), a women's cooperative development organisation, the charity provide targeted assistance to a diverse range of individuals, including prostitutes, artisans, sharecroppers and fisherwomen (Aitken, 1998). As one of their poverty alleviation initiatives, Adithi have revived the production of *sujuni*, a heavily embroidered quilt that is indigenous to the Bihar region (Craftmark, no date). Traditionally gifted to expectant mothers for the swaddling of newborns, *sujuni* are made by stitching together layers of old saris, cloth scraps and dhotis (Strycharz, 2014). The quilt is then decorated with hand-drawn images, which are outlined using multi-coloured chain stitching and filled in with applique and fine running stitches.

When the project first began, the few women who joined endured ridicule but met little resistance (Strycharz, 2014).[9] Many villagers were sceptical of the ability of quilt-making to transform, whilst men allowed their wives to participate because they perceived *sujuni* "as a women's hobby, not a source of income" (Strycharz, 2014: p.136). Nevertheless, as the quilts started to sell, and the women began earning money, the project received greater recognition and participant numbers increased from 5 to 600 women. For a number of individuals, simply attending the fifteen-day training programme was an accomplishment in itself; due to the prevalence of oppressive patriarchal customs, women seldom left their homes and were forced to practice *ghunghat* – the veiling of one's entire face – on the rare occasions that they did (Strycharz,

2014). However, since the introduction of the *sujuni* project, such practices have declined. As wage earners in their own right, women have gained greater independence and are able to move around more freely and interact with men (Craftmark, no date, see also website in note 9). Likewise, programme partici-pants have reported improved household relations, with fewer marital dis-putes and more equitable decision-making (Aitken, 1998). As many of their husbands are unemployed or seasonal labourers, the women's income has provided a small but crucial means of security, lifting families out of mere subsistence. As such, wives are now perceived as less of a burden, allowing some girls to negotiate marriage without a dowry (Aitken, 1998). Moreover, in the villages where Adithi operates, the *sujuni* project has been credited with increasing the age of marriage and improving both numeracy and literacy rates (Strycharz, 2014).

Yet, the newfound respect that the women command is not only derived from their earnings. By moving *sujuni* production out of the home, Adithi has bought together women from different families, paving the way for solidarity and collective action that transcends caste boundaries (Aitken, 1998). Indeed, Tyabji (1999) discusses how the *sujuni* project is not only the revival of a Bihari quilt-making practice, but also its reinterpretation. Instead of traditional religious and pastoral motifs, participants are encouraged to exert creative autonomy and embroider images from their own lives (Strycharz, 2014). The resulting quilts are a vehicle for social commentary, depicting issues such as female infanticide, domestic violence, AIDS transmission, and the harmful effects of pollution (Aitken, 1998; Strycharz, 2014, see also website in note 8). Other *sujuni* display the women performing daily tasks, such as making several trips to the village water pump, collecting bamboo for basket weaving, and cooking over a smoky wood-burning stove. Although less politically overt, these quilts bring to light women's forgotten labour and make a powerful statement about their triple burden. Likewise, as an outlet for participants' fears, frustrations and ambitions, the practice of quilt-making is incredibly therapeutic and emancipatory. Adithi sells the *sujuni* to buyers across the world and foreigners often visit the villages to see the quilts and speak to the women about their work (Aitken, 1998). This provides individuals with increased confidence and self-worth. The group-nature of *sujuni* production also facilitates the discussion and dissemination of ideas. For example, when "the women meet to embroider a quilt on dowry... their conversations revolve around the subject of marriage, dowry, marital crises they have heard about, and their own resolutions for the future" (Aitken, 1998: p.5). Such dialogue has incited women to assemble and fight for their rights, leading to the creation of *panchayats*, namely formal meetings that are held as counterparts to those called by the male village leaders. The decisions made at these gatherings have achieved significant political influence, with the women in one village forcing the closure of the local liquor store, whilst another group successfully advocated for the *sujuni* project's continuation after it was blamed for a love marriage between a Muslim and a Hindu (Aitken, 1998).

The embroidery of *sujuni* can be seen as a form of craftivism, namely "the practice of engaged creativity, especially regarding political or social causes" (Greer, 2007: p.401). Coined in 2003 by American Betsy Greer, it involves practitioners – otherwise known as 'craftivists' – harnessing their creative energy and individual skillsets to help address particular issues (Greer, 2007). As a very personalised practice, the coming together of craft and activism occurs in a variety of different ways, and across a range of settings and locations. For example, some individuals, like the *sujuni* quilters, sew political messages onto fabric, whilst others perform 'knit-ins' or 'stitch-ins' to protest against specific actions, policies or events. Although not restricted to needlework, craftivism has been dominated by practices such as cross-stitch, quilting and knitting as third-wave feminists, particularly within the Global North, attempt to reclaim the domestic arts for feminist expression (Chansky, 2010). As can be seen with the Bihari women, the meditative nature of needlework enables individuals to reflect deeply on injustice and transform their anger and frustration into a powerful, tangible object. Consequently, craftivism has become a preference for many 'burnt-out' campaigners, and those who prefer slower, more gentle forms of activism, because the expression of emotion through craft requires conscious planning, something which, in turn, enables feelings to be conveyed in a more ordered and effective manner (Greer, 2008; Corbett and Housley, 2011). The activist's message may also have greater accessibility, as the association of needlework with both the home and older generations of women means that people approach these crafted objects with interest, rather than automatically rejecting them out of discomfort (Chansky, 2010). As an unexpected political medium, the use of craft for activism only enhances messages of social commentary.

Although the *sujuni* project adheres to the broad definition of craftivism, the Bihari women are, in many ways, unconventional craftivists. For them, quilt-making was not revived for activism, but as an income generation strategy, a means of alleviating women's poverty when stringent social customs prevented other forms of work. Similarly, when programme participants began embroidering political parables, it was not an *alternative* method of voicing discontent, but rather their *only* means of conveying the hardships endured in a highly patriarchal and feudal society. These women also lack many of the resources available to craftivists living in the Global North. As such, they must produce their quilts for sale, yet this has the potential to restrict their choice of subject matter (Aitken, 1998). Indeed, for the average consumer, violent narratives of dowry deaths and female infanticide are not appropriate forms of home decoration. Likewise, the bright threads and fabric preferred by villagers contrast with the muted shades that urban Indian buyers desire for an 'ethnic chic' aesthetic (Aitken, 1998). Due to the steady demand for household products with traditional, picturesque village designs, the project could be sustained with the revenue received from the cushion covers and scarves that are produced alongside the *sujuni*. However, as it is story quilts that the women take interest and pride in (Aitken, 1998), a decline in *sujuni* sales would reduce the project's impact. This demonstrates how market relations, although necessary for alleviating women's economic poverty, can work to the detriment of social empowerment.

Unpicking the politics of sewing

To conclude, neither the sewing machine nor the needle and thread are benign objects. As this chapter has demonstrated, each possesses a distinct material politics, embodying particular forms of power and authority. In each of the case studies discussed, organisations have positioned sewing skills as a means of liberation. However, the extent to which this is being realised is, in reality, highly variable. For example, in Cambodia, sex workers are promised that sewing lessons will transform their lives. Yet after being violently removed from brothels, they are imprisoned in re-training centres and forcibly (re)incorporated into the garment industry. Here they work in dangerous conditions for long hours and little pay, becoming pawns of the anti-trafficking movement, US conservatism, corporate capitalism and neoliberalism. Likewise, Worldwide Proclamation assures its students that sewing skills will enable them to find employment or start their own businesses, bringing both themselves and their families out of poverty. Nevertheless, besides being compelled to receive Christian instruction and perform devotions, there is no evidence that the women receive sufficient training, equipment or opportunities to successfully apply their newly learnt skills. On the other hand, Adithi has succeeded in facilitating the economic and psychological empowerment of its quilters. Through the embroidery of *sujuni*, Bihari women have gained a voice in their households and communities, enabling them to both raise awareness of social ills and reduce their prevalence. However, due to the women's poverty, they are bound into economic relations that may reduce their creative freedom and limit the political potential of quilt-making.

What can be seen is that sewing is not only a tool for the creation of textile items, but also for activism, discipline, the dissemination of ideology and the generation of income – although, as shown by the Cambodian garment industry, not necessarily for the sewers themselves. In this sense, sewing is able to both enhance and remove the agency of women. At the same time, the ways in which sewing has been, and continues to be, harnessed and deployed, is bound up with how different actors perceive women in the Global South. In each of the organisations discussed, there is a shared belief that women need 'saving', either from traffickers, immorality or poverty. Besides the fact that rescue efforts are often unwanted, many of these initiatives actively reproduce the very oppression that they claim to be targeting. Both sex worker rehabilitation initiatives and faith-based organisations are taking advantage of the assumed benignity of sewing apparatus. As such, they are able to indoctrinate and exploit under the guise of humanitarianism. As sewing projects continue to proliferate in development programmes across the world, we need to spend more time 'unpicking' the complex effects of these projects, and consider their interwoven economic, social and cultural impacts.

Notes

1 http://www.afesip.org/ last accessed 26/7/2017
2 http://agapewebsite.org/ last accessed 26/7/2017
3 http://www.christianaid.org/News/2014/mir20140611.aspx last accessed 26/7/2017
4 http://www.christianaid.org/News/2014/mir20140611.aspx last accessed 26/7/2017
5 https://www.mtw.org/stories/details/sewing-for-jesus last accessed 26/7/2017
6 http://www.oacom.org/missions.asp?lid=16 last accessed 26/7/2017
7 Ibid.
8 http://www.paramparaproject.org/institution_adhiti.html last accessed 26/7/2017
9 http://expressindia.indianexpress.com/fe/daily/19990425/fle25049.html last accessed 26/7/2017

References

Aitken, M. (1998) The narrative thread: women's embroidery from rural India, Asia Society Galleries, 9th June to 9th August.

Alvi, S. (2014) *From Sex Worker to Seamstress: The High Cost of Cheap Clothes.* Available at: https://www.vice.com/en_us/article/the-high-cost-of-cheap-clothes-198 (Accessed: 10th April 2017).

Asian Pacific Network of Sex Workers (APNSW) (2017) *Welcome to APNSW.* Available at: https://apnsw.info/ (Accessed: 5th May 2017).

Barry, A. (2013) *Material Politics: Disputes Along the Pipeline.* Chichester: Wiley-Blackwell.

Beaudry, M. C. (2006) *Findings: The Material Culture of Needlework and Sewing.* New Haven: Yale University Press.

Bernstein, E. and Jakobsen, J. R. (2010) *Sex, Secularism, and Religious Influence in U.S. Politics.* Available at: https://www.opendemocracy.net/5050/elizabeth-bernstein-janet-r-jakobsen/sex-secularism-and-religious-influence-in-us-politics (Accessed: 8th May 2017).

Better Factories Cambodia (2017) *The Garment Industry.* Available at: http://betterfactories.org/?page_id=25 (Accessed: 17th May 2017).

BEAPA (Bureau of East Asian and Pacific Affairs) (2016) *U.S. Relations with Cambodia.* Available at: https://www.state.gov/r/pa/ei/bgn/2732.htm (Accessed: 6th May 2017).

Chan Thul, P. (2016) Cambodia raises 2017 minimum wage for textile industry workers. Available at: http://www.reuters.com/article/cambodia-garment-idUSL3N1C51OD (Accessed: 3rd August 2017).

Chansky, R. A. (2010) A stitch in time: third-wave feminist reclamation of needled imagery. *Journal of Popular Culture*, 43(4), pp. 681–700.

Chant, S. (2014) Exploring the "feminisation of poverty" in relation to women's work and home-based enterprise in slums of the Global South, *International Journal of Gender and Entrepreneurship*, 6(3), pp. 296–316.

Christian Aid Missions (2014) *Garments for God's Glory.* Available at: http://www.christianaid.org/News/2014/mir20140611.aspx (Accessed: 12th May 2017).

Clarke, M. (ed.) (2013) *Handbook of Research on Development and Religion.* Cheltenham: Edward Elgar Publishing Limited.

Corbett, S. and Housley, S. (2011) The craftivist collective guide to craftivism, *Utopian Studies*, 22(2), pp. 344–351.

Craftmark (no date) *Sujni Embroidery* [Online]. Available at: http://www.craftmark. org/sites/default/files/P027%20Sujni%20Embroidery.pdf (Accessed: 8th May 2017).

Ditmore, M. H. (2014) "Caught between the tiger and the crocodile": Cambodian sex workers' experiences of structural and physical violence, *Studies in Gender and Sexuality*, 15(1), pp. 22–31.

Domosh, M. (2006) *American Commodities in an Age of Empire*. New York: Routledge.

Dowler, L. and Sharp, J. (2001) A feminist geopolitics?, *Space and Polity*, 5(3), pp. 165–176.

Fair Labour Association (2016) *Cambodia Benchmarks and Charts* [Online]. Available at: http://www.fairlabor.org/sites/default/files/cambodia_benchmarks_and_charts_072816.pdf (Accessed: 3rd August 2017).

Goggin, M. D. (2009) Introduction: Threading women, in Goggin, M. D. and Tobin, B. F. (eds) *Women and the Material Culture of Needlework and Textiles, 1750–1950*. Farnham: Ashgate Publishing Limited, pp. 1–12.

Goggin, M. D. and Tobin, B. F. (eds) (2009) *Women and the Material Culture of Needlework and Textiles, 1750–1950*. Farnham: Ashgate Publishing Limited.

Greer, B. (2007) Craftivism, in Anderson, G. L. and Herr, K. G. (eds) *Encyclopaedia of Activism and Social Justice*. Thousand Oaks: SAGE Publications Inc., pp. 401–402.

Greer, B. (2008) *Knitting for Good! A Guide to Creating Personal, Social and Political Change, Stitch by Stitch*. Boston: Trumpeteer Books.

Higgs, E. T. (2016) Becoming 'multi-racial': the young women's Christian Association in Kenya, 1955–1965, in Glimps, B. J. and Ford, T. (eds) *Gender and Diversity Issues in Religious-Based Institutions and Organizations*. Hershey: IGI Global, pp. 24–50.

Israel, E., Laudari, C. and Simonetti, C. (2008) *HIV Prevention Among Vulnerable Populations: The Pathfinder International Approach*. Watertown: Pathfinder International.

Kanogo, T. (2005) *African Womanhood in Colonial Kenya 1900–1950*. London: James Currey.

Karnani, A. (2009) Romanticising the poor harms the poor, *Journal of International Development*, 21, pp. 76–86.

Labour Behind the Label and Community Legal Education Centre (2013) *Shop 'til they Drop: Fainting and Malnutrition in Garment Workers in Cambodia*. London and Phnom Penh: Labour Behind the Label and Community Legal Education Centre.

Larsson, E. (2016) Poverty drives wages down. Available at: https://arbetet.se/2017/05/08/poverty-drives-wages-down/ (Accessed: 3rd August 2017).

Mallampalli, C. (2004) *Christians and Public Life in Colonial South India: Contending with Marginality*. London and New York: RoutledgeCurzon.

Microfinance Opportunities (2017) *Will the New Minimum Wage in Cambodia Make a Difference for Garment Workers?* Available at: http://fashionrevolution.org/reality-check/ (Accessed: 3rd August 2017).

Mokosso, H. E. (2007) *American Evangelical Enterprise in Africa: The Case of the United Presbyterian Mission in Cameroun, 1879–1957*. New York: Peter Lang Publishing Inc.

Moore, A. E. (2016) *Threadbare: Clothes, Sex and Trafficking*. Portland: Microcosm Publishing.

Overs, C. (2009) *Caught Between the Tiger and the Crocodile: The Campaign to Suppress Human Trafficking and Sexual Exploitation in Cambodia* [Online]. Available at: http://www.plri.org/sites/plri.org/files/caught-between-the-tiger-and-the-crocodile. pdf (Accessed: 8th May 2017).

Parker, R. (1984) *The Subversive Stitch: Embroidery and the Making of the Feminine.* London: Women's Press.

Parker, R. (2010) *The Subversive Stitch: Embroidery and the Making of the Feminine.* New York: I.B. Tauris & Co. Ltd.

Pearson, R. (2007) Reassessing paid work and women's empowerment: lessons from the global economy, in Cornwall, A., Harrison, E. and Whitehead, A. (eds) *Feminisms in Development: Contradictions, Contestations and Challenges.* London: Zed Books, pp. 201–213.

Prentice, R. (2017) Microenterprise development, industrial labour and the seductions of precarity, *Critique of Anthropology*, 37(2), pp. 201–222.

Rehman, M. (2010) Muslim politics in India and the 15th General Elections, in Mehra, A. K. (ed.) *Emerging Trends in Indian Politics: The 15th General Election.* New Dehli: Routledge, pp. 133–156.

Robert, D. L. (2008) The 'Christian Home' as a cornerstone of Anglo-America missionary thought and practice, in Robert, D. L. (ed.) *Converting Colonialism: Visions and Realities in Mission History, 1706–1914.* Grand Rapids: William B. Eerdmans Publishing Company, pp. 134–165.

Rose, N. (1999) *Powers of Freedom: Reframing Political Thought.* Cambridge: Cambridge University Press.

Schwittay, A. (2011) The marketization of poverty, *Current Anthropology*, 52(S3), pp. S71–S82.

Siddiqi, D. (2009) Do Bangladeshi factory workers need saving? Sisterhood in the post-sweatshop era, *Feminist Review*, 91, pp. 154–174

Smirl, L. (2011) *Drive by Development: The Role of the SUV in International Humanitarian Assistance.* Unpublished paper.

Strycharz, H. (2014) Sewing voices: the *arpilleristas* and the women of the Adithi Collective, in Greer, B. (ed.) *Craftivism: The Art of Craft and Activism.* Vancouver: Arsenal Pulp Press, pp. 133–140.

ter Haar, G. (ed.) (2011) *Religion and Development: Ways of Transforming the World.* New York: Columbia University Press.

Thomas, N. (2002) Colonizing cloth: interpreting the material culture of nineteenth-century Oceania', in Lyons, C. L. and Papadopoulos, J. K. (eds) *The Archaeology of Colonialism.* Los Angeles: Getty Research Institute, pp. 182–198.

Tomalin, E. (2013) *Religions and Development.* London and New York: Routledge.

Tyabji, L. (1999) *The Story Behind the Stitches: Indian Women, Indian Embroideries.* Available at: http://www.textilemuseum.ca/exhibitions/past-exhibitions/exhibition-essays/stitching-women%E2%80%99s-lives-sujuni-and-khatwa-from-bih (Accessed: 5th May 2017).

US Department of State (2014) *Trafficking in Persons Report* [Online]. Available at: https://www.state.gov/documents/organization/226844.pdf (Accessed: 8th May 2017).

US Press Secretary (2003) *Trafficking in Persons National Security Presidential Directive.* Available at: https://2001-2009.state.gov/g/tip/rls/other/17966.htm (Accessed: 11th May 2017).

Winn, P. (2016) *Why Cambodia's Sex Workers Don't Need to be Saved.* Available at: https://www.pri.org/stories/2016-03-29/why-cambodias-sex-workers-dont-need-be-sa ved (Accessed: 9th May 2017).

Worldwide Proclamation (2015) *India OAC Sewing Schools Report, August 2015.* Available at: http://www.oacom.org/missions.asp?lid=16 (Accessed: 14th May 2017).

Worldwide Proclamation (2017) *India- Sewing Schools.* Available at: http://www.oa com.org/missions.asp?pid=16 (Accessed: 14th May 2017).

Zimmerman, Y. C. (2013) *Other Dreams of Freedom: Religion, Sex and Human Trafficking.* Oxford: Oxford University Press.

6 Work, value and space

Three key questions of making for the Anthropocene

Chantel Carr, Chris Gibson, Elyse Stanes and Andrew Warren

Introduction

Debates about the precarious ecological state of the world are gathering pace. The Anthropocene is a term that has been proposed across the natural and social sciences to describe a new geological epoch defined by the unprecedented influence of humans on climate and the environment (Castree, 2014; Head, 2016). Making, where it is connected with the carbon-intensive transformation of raw materials into usable things, arguably sits at the centre of Anthropocenic concerns (Bridge, 2001, 2009; Castree, 2004; Cook, 2004). The question of making, under such conditions, is increasingly one of sustaining everyday life in a way that does not demand abundant resources. Initiatives such as the Circular Economy (Hobson, 2016), the field of industrial ecology and investigations into product stewardship (Lane and Watson, 2012) offer a range of approaches to the environmental challenges that accompany making *at scale.*

From a different perspective but equally relevant in the context of the Anthropocene, are those activities aligned with cultures of re-making. The skilled work of repair, restoration, recycling, rebuilding, rehabilitation and renovation is inherently tied to how societies might materially *respond* to destabilising forces (Carr, 2017; Gregson, Metcalfe & Crewe, 2009; DeSilvey, 2007). The Anthropocene thus raises questions around what kinds of economies will become necessary and even desirable, in a future characterised by volatile weather events, ecosystem disruptions and resource scarcities (Hobson, 2016; Gibson and Warren, 2016). It brings into question a capital-centric view that values growth and the making and selling of new things at ever-increasing rates. Moreover, the Anthropocene challenges scholars to critically engage with how particular 'making' bodies are being devalued at a time when their skills could be deemed most relevant.

As critical geographers with an interest in how particular economic narratives are constructed and re-circulated (Massey and Rustin, 2014), we have sought to understand how the Anthropocene and its destabilising forces intersect with cultures of making across a diverse range of projects. We are four researchers with distinctive but sometimes overlapping concerns, informed variously by Marxian, feminist and poststructuralist thinking. A common thread to our work is to develop an understanding of how making and material skill might still come to matter (indeed we argue that it will matter more than ever) in the context of mounting environmental and economic uncertainty (Carr and Gibson, 2016; cf. Hudson, 2012). From the hulking blast furnaces of steelmaking to emergent makerspaces, and from guitar lutherie and surfboard-shaping workshops to clothing fabricultures, we have explored how different cultures of making might foster dispositions oriented toward living more thoughtfully with materials and technologies.

In this chapter we come together to take up three questions that examine themes of value and the *work* of the Anthropocene, through the empirical lens of making. These themes continue to develop and coalesce in ad-hoc conversations as we each move iteratively (and sometimes collaboratively) between the field and the campus. The questions are not prescriptive, but rather are intended to seek out points of intersection across divergent projects: around how particular tasks and skills associated with making are valued according to their spatial and temporal configuration, and their relationship with hegemonic capitalist economies focused on accumulation and profit. Our focus here is making within and beyond the enterprise, as a means of connecting broad critiques of production and consumption with discourses of environmental crisis. In each of the three questions that follow, we seek to ground these themes with concrete empirical examples. Ultimately, we contend that responses to the Anthropocene must first come to terms with capitalism's labours, how such labours are valued and how they might be valued differently, within a more catastrophic ecological epoch.

Maria Puig de la Bellacasa (2011) has argued for rethinking relations of care within a framework of already existing materials and technologies. We share this sentiment, but extend it to focus on the already existing labours that are enrolled in making and remaking things. In the Anthropocene, the conditions for survival are ultimately social, and raise questions of resourcefulness in everyday life (Gibson, Head and Carr, 2015). We seek to demonstrate that those who are skilled at working physically with materials are already doing this work within formal enterprises, informal and ad hoc economies, communities and homes. Many such workers possess dispositions and ingenuities that will undoubtedly be needed in a future characterised by more volatile ecological and economic conditions. Exactly how these skills accrue, how they are valued or under threat, and how they spill from and across spaces of formal production, informal economy and domestic work, are far from mere academic questions about labour in the present day. They are future-oriented questions at the heart of societal capacities to respond to unanticipated disruption ahead, and for this reason, are our focus in this chapter.

Jobs or tasks?

The first question we pose is simply *jobs or tasks?* This inherently scalar question may seem obvious to many of our colleagues, including those in this volume, who work at the scale of the body in action. But where our work has intersected with scholarship on economy and labour, we have observed a lacuna in interest concerning what people *do* at work. This is particularly acute for those who 'make' at the scale of the capitalist enterprise, where debates about the future of manufacturing centred on robotics and automation remain curiously opaque as to how such work is composed (Dunlop, 2016; Srnicek and Williams, 2015). The nature of making *as work* has shifted dramatically over the last two hundred years. Within the last half century alone, automation, global divisions of labour, and the rise of the information economy have radically reconfigured how making proceeds, and where. Yet few accounts of the changing nature of material work have examined what workers do all day, every day – their physical tasks and embodied labour processes. In the context of the Anthropocene, what matters is not just how the labour market is being reconfigured by macro-economic processes, but how capitalist, noncapitalist and informal economic systems foster the skills and dispositions to connect with, and to live more thoughtfully amongst, materials and things. Tasks then, are a critical scale of analysis in understanding what embodied capacities for working skilfully with materials lie within existing maker cultures.

A host of technologies, including robotics and additive manufacturing, are redefining manufacturing processes and widening possibilities for what can be made, and where (Birtchnell and Urry, 2013; Bissell and del Casino Jr, 2016). Across industries such as surfboard-making and guitar manufacture, we have traced the impact of automation on how things are made. In these industries, while social impacts are surely felt in terms of job losses (where they occur), a focus on tasks draws to light questions of how particular skills, input materials, and labour processes are being reconfigured, rather than eviscerated. Some material knowledges and skills – of rare timbers or local wave conditions for example – have re-emerged as more significant. These details ultimately connect with – and indeed shape – future resourceful capacities in the context of the Anthropocene.

In surfboard making, labour process has shifted due to automation, but not without reference to existing labour process (Warren and Gibson, 2014). A hand-based method of production dominated the market until relatively recently. Historically, surfboards were ordered from local workshops and made entirely by hand, the exact cut and shape tailored to the relation between an individual surfer's body, weight, height, surfing style, and prevailing local wave types and size. Two main tasks, and forms of work, prevailed: shapers, who carve boards into final shape, and glassers, who seal the boards to protect them and create buoyancy. In the 1990s technology enabled the development of automated cutting of blanks. Computer-numerical control (CNC) machines began to proliferate, workshops up-scaled production volumes and sought export markets. The traditional work of shaping appeared doomed. Yet jobs did not simply disappear in a causal sense. Rather, tasks were reconfigured, and not without contestation. Hand-shapers instead became

R&D experts, developing prototypes that could be programmed into CNC machines. New tasks arose that created work in programming; while glassing and sanding tasks were still needed. More recently, a shift in the cultural capital of iconic makers in surf magazines, blogs and websites has reinvigorated appreciation among surfers for local board making. The sum of this is the emergence of a hybrid surfboard-making workshop: where makers design and shape original designs, offer higher-end custom services to expert surfers in the 'traditional' method, but also fabricate and sell boards cut by machines, but still sanded and glassed by hand, typically to beginner and intermediate surfers.

Meanwhile, within guitar manufacturing, the advent of CNC machines has replaced some highly skilled craft tasks of cutting and carving guitar necks by hand (Dudley, 2014). Yet reducing labour or simplifying tasks was not especially the goal. Greater use of CNC machines was driven by a concern to increase the consistency and accuracy of product, but also to address health and safety concerns stemming from timber-cutting, sawdust and toxic chemicals. As one Australian manufacturer interviewed explained:

> We've been on CNCs now for 23, 24 years. It's given us much greater consistency in terms of dimensional accuracy specifications. It's removed a lot of work that was potentially dangerous... When I first started, there were a number of guys with missing fingers on the staff. There aren't any, now.

Indeed, labour intensity persists for tasks machines cannot perform, especially in relation to the properties of component timber materials, and skills needed to 'read' and manipulate them. In guitar-making the traditional timbers used are now scarce or heavily regulated (both conditions emblematic of the Anthropocene), and the industry is adjusting to new alternative species and sources of salvaged timbers (Gibson and Warren, 2016). Tasks such as 'reading' incoming supplies of logs and billets for inconsistences and planning to maximise yield from them, new techniques for drying wood, and sanding for imperfections, have taken on heightened importance. As the above manufacturer explained:

> Guitars are appalling for mechanisation. There's hardly a flat surface on them, and if there is you can't touch it with anything because it dents. There's hand work everywhere. I don't see that changing. Even the mass producers, they're just using more people doing hand work, really. You can use a machine to improve your accuracy and consistency. Great. I don't see handmade or hand-built guitars going away, not unless we change from wood.

Rather than suffer 'inevitable' replacement by machines, in an era of greater Anthropocenic uncertainty around timber raw materials, workers with gradually acquired material knowledges and haptic skills with wood accrue a degree of agency within an evolving labour process, and within concrete spaces of work. For geographers concerned with making, then, it is critical to look beyond generalised macroeconomic predictions of people being replaced by machines or robots, to the retention and reconfiguration of specific tasks, materials and skills in a context of growing ecological uncertainty.

How are different modes of making valued?

The second question we pose reveals our concern with how particular making tasks are valued according to how they are configured in relation to dominant modes of production. Our empirical investigations point to an historically entrenched overemphasis on the processes by which commodities are *newly* produced, and a corresponding lack of emphasis on those skills and tasks that either underpin the production of commodities, or which sustain their existence in the world. Important insights into where resourceful capacities for the Anthropocene might be located only emerge when these labours are momentarily disentangled from those cultures of making that are concerned with the production of new things. The issue is one of visibility. Where cultures of making are focused on sustaining things that already exist in the world, the mundane work of caring for things and materials, including cleaning, adjusting, maintaining, repairing and repurposing is largely invisible (see DeSilvey, Ryan and Bond in this volume). Though the work associated with sustaining commodities in the world may involve waged labour, it is work that often enrols and fosters 'broad' skills that are difficult to capture in more conventional analyses, including those focused on jobs rather than tasks.

To illustrate, we draw from an ethnography of maintenance and repair workers within a large steel plant in our home city of Wollongong (Carr, 2017). Steelmaking is, in many ways, the archetype of modern industrial maker cultures, with enormous plant, complex equipment and unfathomable temperatures. The central place of steelmaking in both academic and popular narratives of the industrial revolution, and subsequent rust belt decay, has reified this form of making: 'the steelworker' is a motif of modernity. Though rarely have such accounts distinguished between the work of steel*making* – which for at least the last 50 years has been largely disembodied – and the litany of work that keeps a steel plant operational.

It should be rather obvious in light of advances in manufacturing throughout the twentieth century, but no one actually *makes* steel. That is, within the labour processes of steel production, workers do not physically handle the product that is made. In a big industrial complex dominated by materials that are made pliable through extreme heat (at the height of the process, molten steel reaches 1200°C), it is machinery that heats, mixes, squeezes, rolls, cools and moves the product in its various states. Clusters of instruments feed data about the process to control systems, which in turn are monitored by production workers. Contrary to stereotypical depictions of steelmaking, these workers are highly skilled, and often come from para-professional backgrounds in metallurgy or from skilled trades. But their skill is largely directed toward screens that supply data about the process.

Yet outside the control room, there are other kinds of embodied *material* work that are fundamental to how a steel plant operates. Maintenance and repair is the smoothing work that restores order and function, as the hot and volatile process of steelmaking takes its toll on machinery. As yields rise and fall, such work is unremitting, yet it is imperceptible in yield figures. The effect is one of deep imbalance in how the different labour cultures of the industrial site have come to be understood, where the modes of making that underwrite production are seldom seen. Subsuming these cultures within discourses of production is a significant oversight (Carr, 2017). As one participant explained:

> And people don't seem to realise that it's a continuous process, and if one limit switch fails, it stops. The whole mill stops and ... that's what they say it costs up to ... well then ... fifteen hundred dollars a minute, depending on what they're rolling, so the pressure was on you to get that mill going as quickly as possible.

Where workers are responsible for making running repairs to the machinery that makes steel, they become skilful at responding rapidly and creatively across scales and technologies. Other workers are part of large maintenance teams, who clean, adjust, repair, rework and reconfigure a vast array of machinery, equipment and technologies that support the quotidian production process. These contingent (re)maker cultures – embedded deep within the modernist industrial complex – foster a range of skills and dispositions that are precisely oriented towards the kind of uncertainty that is being associated with the Anthropocene. In the following section, we explore how these cultures move beyond the bounds of paid work, and into the home and community.

Fashion and clothing fabricultures are a second example where there is an emphasis on the processes by which commodities are produced, while those skills and processes that sustain the existence of such commodities in the world are notably less explored. Fusing livelihoods and labour, creative expression, fundamental and utilitarian need and personal pleasure, fashion is at the heart of contemporary maker culture (Fletcher, 2016). Yet within academic work focused on the fashion sector, the moral geographies of production remain prominent, led by a necessary focus on exploitative labour (McRobbie, 1997, 1998; Rantisi, 2014; Braitch and Brush, 2011; Hall and Jayne, 2015). Meanwhile, scholarship on fashion consumption has focused overwhelmingly on relationships between the body and identity in consumer culture. Fashion consumers are regularly appraised for eschewing environmental and social responsibilities.

Yet the informal and mundane skills, tools, labour and material transformations that are essential to understanding consumer connections with material things are often overlooked. Research with fashion consumers has revealed a range of tasks that are involved in the cyclic rhythms of wearing and caring for clothing. Varied accounts of material resourcefulness, frugality and thrift craft small lessons of change in new economic, environmental and socially sustainable modes of consumption (Gregson and Crewe, 1998, 2003; Hitchings, Collins and Day, 2015). Individual haptic skills, collective cultural norms and the agency of products themselves shape the value and ongoing care and maintenance for things. Following this approach offers opportunities to discover how clothing is valued, used, cared for and maintained (Gill, Lopes and Kaye-Smith, 2016; Fletcher, 2016) – and to follow a different set of spatialities and temporalities than those characterised by fast and excessive consumption.

An ethnography of the 'fashion journeys' of young adults in Sydney, Australia (Stanes, 2018) revealed numerous ways that clothing is valued, cared for and looked after. Regular acts of mending were notably absent from everyday routines of clothing maintenance. Instead, care for clothes came through a set of mundane haptic practices that sustained the life of clothes so that mending was unnecessary. Lauren, for instance, preserved delicate garments 'that basically disintegrate with wear' in repurposed shoe boxes that fit neatly into the underutilised shelves in her 1960s chipboard wardrobe, thus saving the garment from damage. Drawing on lessons learnt from his Filipino parents, Felipe enacted a particular set of laundry practices for caring for intimate garments – especially underwear:

> you wash them with hot water but you wouldn't wash the garter, you know? The rubber bit. And um, after you wash, after you wash that you wouldn't dry it in direct sunlight so that the fabric and the garter don't deteriorate.

Bede proudly bragged about the 'pretty much vintage' pair of jeans – still in regular use, and worn over the past 10 years since he was 12 years old: 'I guess I tend to hold onto clothing because of how well I maintain it and because of how much I value it. I'm very sentimental like that'. There were many more examples of Bede's garments that were worn with care – including trainers, shirts and jumpers – that were not expensive, nor of great quality, but that persisted because of the care taken to preserve them. Trainers, for instance, were always cleaned and kept in a box after use. Bede concedes that he doesn't feel that he cares for his clothes in a particularly noteworthy way. Rather, he *cared about* his clothes.

Attending to the informal and mundane skills that speak to care and sustainment of garments – sensing out quality, mending, sharing/hand-me-downs, laundry and storage – point to a range of skills that have value beyond the monetary. Such acts are in effect work tasks that sustain the material coherence and the use value of clothing, though rarely are these conceptualised as work in academic or policy discourse. Instead, they are typically called 'practices' in the sphere of 'consumption', cast in opposition to the waged making work of the assembly line or garment design. Beyond the aura of the finished com-modity, and analogous to industrial maintenance tasks, these are undervalued but enormously useful labours, that under Anthropocenic conditions can be connected with value through extending the practical lives of material things.

How is the value of making configured spatially?

The final question we ask is specifically concerned with the *geographies* of making, and how making work comes to be differentially valued according to where it proceeds. At the heart of this question lies a persistent and of course acutely modernist dichotomy, which works to divide spaces of 'leisure' (includ-ing the home) from those of 'work'. Feminist geographers have consistently and persuasively argued that such dichotomies frame dominant systems of valuing and devaluing work, while overlooking the ways in which paid and unpaid tasks might intersect (England and Lawson, 2004; Hanson and Pratt, 1988; Mackenzie and Rose 1983; McDowell 1999). In an Anthropocenic light, exactly which skills, tasks and dispositions emerge to be 'counted' compels attending more closely to spatial and gendered narratives, and the ways in which many cultures of making have always skirted the edges of community and capitalist spheres.

The cleaving apart of home and work is intimately bound up with the history of industrial production, and so too then with the history of making (Gough, 2003). For much of the nineteenth and twentieth centuries, ideals of work as waged labour in a formal setting outside the home were tied to the masculine identity of the factory worker (McDowell, 1999). The 'feminine' home was constructed in opposition as the place 'where the physical and emotional maintenance of the worker was carried on, readying him to go back to work in the morning, clothed, fed, satisfied and relatively willing to put in an efficient day's work on the production line' (Mackenzie and Rose, 1983: p.157). The idea of men leaving the home to work and returning to its 'haven' at the end of the shift was disrupted in the 1970s, with a transition to service-based economies. A 'crisis of masculinity' emerged from the decline in jobs marked by physical work and the production of tangible things, and the erosion of solidarity in working-class communities built on manufacturing legacies (McDowell, 2003). Women's participation in paid employment outside the home rose significantly, though the work was heavily weighted towards those qualities that have always been regarded as 'women's work': care, empathy, nurturing and sociality (McDowell, 1999, 2015).

Yet some cultures of making have offered new perspectives on these normative spatial and gendered framing of value and work. We have found qualities such as care evident in the making cultures of industrial repair and maintenance, as well as emotional labours in unexpected settings like surfboard manu-facturing workshops, which are overwhelmingly masculinised (Warren and Gibson, 2014). We have also traced different making cultures to origins in the home, backyard and garden, where they do valuable work in forming and maintaining social communities, as well as contributing to material economies of thrift and making-do. Older cultures of surfing manufacture and steelmaking cultures of skilled repair and maintenance for example, along with burgeoning contemporary forms of home-based craft work (supported by technology) frequently transgress the bounds of home and work as well as paid and unpaid modes. In the process they offer fruitful models for establishing new connections between existing skills and dispositions, and new ways of living. Under such conditions, the activities of 'modernist' manufacturing cultures have the capacity to make space for experiment, uncertainty and fragility (Gibson-Graham, 2011). Such work catalyses different notions of value when connected with Anthropocenic conditions characterised by scarce resources.

To illustrate how making labour is valued differently depending on where it proceeds, we return to the example of steelworkers and the industrial maker cultures of our home region. There is more to this story of industrial repair and maintenance than a simple lament for unrecognised skill in the paid workplace. The skills and dispositions accrued within large-scale industrial enterprises like Port Kembla also travel home via the bodies in which they accumulate, where they serve different functions (Carr, 2017). Participants derive satisfaction and enjoyment from the challenges of their labour, such that they commonly deploy the same maintenance and repair skills, methods and ethics in their 'own' time. These trades-qualified workers frequently fit new power outlets and fix power tools for themselves and family members. They experiment with plumbing systems and build chicken sheds from salvaged – and often sophisticated – materials. They design and fabricate bespoke barbeques and wood stoves from valuable materials like stainless steel, offcuts gleaned from the excesses historically encountered in paid work. They 'hack' everyday objects from bar stools and garden spades to hot water systems and solar panels, in order to improve on the designs of commodity manufacturers, or to adapt them to the micro-spaces and climates of their surroundings.

This work is deeply embedded within existing networks of reciprocity that constitute everyday life both within and outside the enterprise. The social networks that have emerged through generations of apprenticeship and inculcation in this city's industrial maker culture are also evident outside of paid work. Workers are both thrifty and generous. They are assiduous sharers but also motivated to live autonomously. They demonstrate resourceful and inherently *productive* ways of living in the suburbs – a setting more often maligned for its homogenous, consumptive cultures. Acknowledging that particular modes of work resonate across the bounds of paid and unpaid activity as well as industrial and domestic space creates an opportunity to re-evaluate the skills and work dispositions that industrial workers embody, as well as to re-visit persistent assumptions about the dualisms of work and leisure in the industrial city. Skills are directed simultaneously towards capitalist extraction of surplus value, but also the sustainment of everyday life. One participant for example, described the frustration he felt towards material longevity in consumer appliances, before going on to talk about how he was able to provide a repair service for his family and others in his community:

> And it stems too, a lot of people don't have a trade, or aren't … manually minded in fixing things now … It's, you know, as people say, we're in a throwaway society … And some of the manufacturers are at fault too, because they'll sell you something cheap, but you buy parts for it … good luck.

In the context of the Anthropocene, this example points to deep problem-solving and diagnosis skills developed as legacies of industrial-scale production and training models, but which are also extended beyond the bounds of paid work. Contrary to common depictions of industrial manual work as a trial to be 'endured' during shift, these skills spill out beyond the perimeter of the plant, because they are enfolded into working 'bodies' who also inhabit other spaces. Although undervalued within the restructuring capitalist enterprise, such workers express dispositions towards materials and systems that have already proven to be enormously valuable socially and economically beyond the plant.

While industrial cultures of repair and maintenance provide one perspective on new connections between the normatively separate realms of leisure and work, the production of cultural products such as guitars, surfboards and fabricultures can also be traced to the home. Technological, material and spatial changes within both homes and manufacturing settings have become vital in the transformation and democratisation of craft based industries that draw on skills and knowledge developed in the context of everyday domestic life (Braitch and Brush, 2011; Hall and Jayne, 2015; Richards, 2008). For some time now, an emergent suite of concerns around worker exploitation, the sexist marketing of fashion, and the environmental consequences of throwaway consumerism, has led to an imagining of new possibilities in fashion production, an industry centred on the experiences of women. The fetishisation of 'the material, the tactile, the analogue' alongside the technological affordances of online market places, has seen women's home-based crafts elevated 'from local markets to the high street' (Luckman, 2013: p.251). This renaissance in fashioning goods in spaces such as the home and community is underpinned by an interest in the creation of objects at a 'speed at odds with the imperative towards hyper-production' (Braitch and Brush, 2011: p.236. See also Hall and Jayne, 2015). Contemporary forms of fashion fabriculture have become a means to critique capitalism and its exploitative supply chains and labour practices, a way of inserting agency, power and creativity into making clothes, and a method through which alternative identities and communities are forged (Hall and Jayne, 2015; Luckman, 2013; Fletcher, 2016; Luckman, 2013). In private homes as well as community spaces, the revival of fabricultures, such as knitting and dressmaking have been reimaged through informal and formal groups that encompass leisure and friendship – but also as social and political tools (Hall and Jayne, 2015). Home-based fabricultures are thus being increasingly connected with a moral ethics of environmental sustainability, thrift, and a material awareness of resource scarcity (Fletcher, 2016; Gill, Lopes and Kaye-Smith, 2016; Hall and Jayne, 2015).

One final perspective on the differentially gendered and spatialised work of making comes from surfboard manufacturing. Surfing subculture is a US$10 billion a year global industry, though its origins are embedded in DIY craft work performed in coastal suburban tool sheds and organised around day jobs. Corresponding to other capitalist industries, jobs in the surfboard industry 'are not gender neutral; they are created as suitable for particular sexed bodies' (McDowell, 1997: p.25). Since the late 1950s, commercial surfboard-making has been organised around a manual labour process. Jobs accordingly mapped onto male working bodies (McDowell, 2011). Surfboard-makers talked about their work as tough and physical. On occasion this spilt over into proclamations positioning this work as 'men's work', buying into ideas about strength and suitability of gendered bodies to undertake manual tasks.

On deeper investigation, the veneer of masculinity proved precarious (Warren, 2016). Men talked about bodies falling apart, lung diseases from chemical fumes and sawdust, and the emotional toll of increasingly insecure work. Yet asked to describe the pleasures of the work, they described an emotional terrain of labour guided less by performances of masculine strength and skill, but rather by the social relationships forged within workspaces. Performances of work and commercial transactions in surfboard manufacturing are flushed with close, personal interaction; not only between workers, but also through engagements with customers. Surfboard-makers continue to make boards by hand not because it is especially profitable, but rather because of the satisfaction of making useful things that others love and cherish, for the lifestyle it enables, the relationships it forges, for reasons of emotion, passion, and care (cf. Delyser and Greenstein, 2017). Such dispositions and motivations, just like those expressed by knitters and steelworkers, offer new ways of conceiving of and conceptualising value in light of impending Anthropocenic upheaval.

Conclusion

In policy and technical worlds, making (even, or perhaps especially where it is writ large as manufacturing) is dominated by the idea of creation: its practices exist to bring something into existence, from which profit can be earnt. But a world controlled by corporate interests that values growth, innovation, making and selling new things at ever-increasing rates cannot be sustained. At some point, if not already, the value of those who make things differently, who sustain systems, who repair and re-make material things already with us, will come to the fore. In the meantime, processes that de-value manual skill, material iteration and the sustaining of systems risk jettisoning the very bodies, dispositions and systems of apprenticeship and skills acquisitions needed for a volatile future. Here, we have sought to highlight three questions where the tasks of making and maintaining intersected in complex geometries with valuation processes, gender norms, and geographic spaces of formal enterprises, and beyond into communities and homes.

Amidst the polemic debate about automation and the future of work, mounting concerns about jobs overlooks just what tasks will need attending to. Under Anthropocenic conditions these will likely include things like servicing infrastructure, disaster relief, or rebuilding. At question is not only the power of global capital or processes of restructuring, automation and deskilling that de-value skills and dispositions within the capitalist enterprise, but a wider concern that the need for material labour will persist, and indeed grow, as ecological and social conditions become more calamitous. For these reasons, drawing to light the value of labour beyond profit, masculine skill and iconic workplaces involves being able to recognise and nurture the (otherwise devalued) skills involved in sustaining things, in light of more volatile futures.

Simultaneously, we will need to figure out how to re-value tasks and skills in ways that transcend gender categories and power relations. Like all work cultures, making intimately maps onto established perceptions of gendered bodies and labour value. In a more volatile world, who will do the work of making and who will do the work of sustaining? Where do these skills and cultures currently reside? The challenge we face as scholars interested in making is to break outside entrenched positions, and to link careful analyses of microspaces and actors to broader debates about gender, climate change, economic collapse and capacities to cope with and adjust to extremity. Across diverse modes of production – craft-based, assembly line, maintenance, repair – and across diverse spaces – artisanal workshops, steelmills, homes, makerspaces – lie skills and dispositions that open new possibilities. Making is a critical and complex part of the discussion about how we connect the (over)-production of stuff with the climate change discourse, and how we comprehend alternatives within the exigencies of everyday life and work.

References

Birtchnell, T. and Urry, J. (2013) Fabricating futures and the movement of objects. *Mobilities*, 8, 388–405.

Bissell, D. and del Casino Jr, V. (2016) Wither labor geography and the rise of the robots? *Social & Cultural Geography*. 18(3), 435–442.

Braitch, J.Z. and Brush, H.M. (2011) Fabricating activism. *Utopian Studies*, 22(2), 233–260.

Bridge, G. (2001) Resource triumphalism: Postindustrial narratives of primary commodity production. *Environment and Planning A*, 33(12), 2149–2173.

Bridge, G. (2009) Material worlds: Natural resources, resource geography and the material economy. *Geography Compass*, 3(3), 1217–1244.

Carr, C. (2017) Maintenance and repair beyond the perimeter of the plant: Linking industrial labour and the home. *Transactions of the Institute of British Geographers.* 42(4), 642–654.

Carr, C. and Gibson, C. (2016) Geographies of making: Rethinking materials and skills for volatile futures. *Progress in Human Geography*, 40(3), 297–315.

Castree, N. (2004) The geographical lives of commodities: Problems of analysis and critique. *Social & Cultural Geography*, 5, 21–35.

Castree, N. (2014) The anthropocene and geography I: The back story. *Geography Compass*, 8(7), 436–449.

Cook, I. (2004) Follow the thing: Papaya. *Antipode*, 36(4), 642–664.

DeLyser, D. and Greenstein, P. (2017) The devotions of restoration: Enthusiasm, materiality, and making three "Indian Motocycles" new. *Annals of the Association of American Geographers*, 0(0), 1–18.

DeSilvey, C. (2007) Salvage memory: Constellating material histories on a hard-scrabble homestead. *Cultural Geographies*, 14, 401–424.

Dudley, K.M. (2014) *Guitar Makers: The Endurance of Artisanal Values in North America*. Chicago: University of Chicago Press.

Dunlop, T. (2016) *Why the Future is Workless*. Sydney: NewSouth Publishing.

Fletcher, K. (2016) *The Craft of Use*. London: Routledge.

England, K. and Lawson, V. (2004) Feminist analyses of work: rethinking the boundaries, gendering and spatiality of work. In *A Companion to Feminist Geography*. Nelson, L. and Seager, J. (eds). Maldon: Wiley-Blackwell, 77–92.

Gibson, C., Head, L. and Carr, C. (2015) From incremental change to radical disjuncture: Rethinking everyday household sustainability practices as survival skills. *Annals of the Association of American Geographers*, 105(2), 416–424.

Gibson, C. and Warren, A. (2016) Resource-sensitive global production networks: Reconfigured geographies of timber and acoustic guitar manufacturing. *Economic Geography*, 92(4), 430–454.

Gibson-Graham, J.K. (2011) A feminist project of belonging for the Anthropocene. *Gender, Place and Culture*, 18(1), 1–21.

Gill, A., Lopes, A.M. and Kaye-Smith, H. (2016) Practicing sustainability: Illuminating 'use' in wearing clothes. *Cultural Studies Review*, 22(1), 32–58.

Gough, J. (2003) *Work, Locality and the Rhythms of Capital*. London: Continuum.

Gregson, N. and Crewe, L. (1998) Dusting down second hand rose: Gendered identities and the world of second hand goods in the space of the car boot sale. *Gender, Place and Culture*, 5(1), 77–100.

Gregson, N. and Crewe, L. (2003) *Second Hand Cultures*. London: Berg.

GregsonN., Metcalfe, A. and Crewe, L. (2009) Practices of object maintenance and repair: How consumers attend to consumer objects within the home. *Journal of Consumer Culture*, 9, 248–272.

Hall, S.M. and Jayne, M. (2015) Make, mend and befriend: Geographies of austerity, crafting and friendship in contemporary cultures of dressmaking in the UK. *Gender, Place and Culture*, 23(2), 216–234.

Hanson, S. and Pratt, G. (1988) Spatial dimensions of the gender division of labor in a local labor market. *Urban Geography*, 6, 367–378.

Head, L. (2016) *Hope and Grief in the Anthropocene*. London and New York: Routledge.

Hitchings, R., Collins, R. and Day, R. (2015) Inadvertent environmentalism and the action–value opportunity: reflections from studies at both ends of the generational spectrum. *Local Environment*, 20(3), 369–385.

Hobson, K. (2016) Closing the loop or squaring the circle? Locating generative spaces for the circular economy. *Progress in Human Geography*, 40(1), 88–104.

Hudson, R. (2012) Critical political economy and material transformation. *New Political Economy*, 17(4), 373–397.

Lane, R. and Watson, M. (2012) Stewardship of things: The radical potential of product stewardship for re-framing responsibilities and relationships to products and materials. *Geoforum*, 43(6), 1254.

Luckman, S. (2013) The aura of the analogue in a digital age: Women's crafts, creative markets and home-based labour after Etsy. *Cultural Studies Review*, 19(1), 249–270.

Mackenzie, S. and Rose, D. (1983) Industrial change, the domestic economy and home life. In *Redundant Spaces in Cities and Regions? Studies in Industrial Decline and Social Change*. Anderson, J., Duncan, S. and Hudson, R. (eds). London: Academic Press, 155–200.

Massey, D. and Rustin, M. (2014) Whose economy? Reframing the debate. In *After Neoliberalism? The Kilburn Manifesto*. Hall, S., Massey, D. and Rustin, M. (eds). London: Soundings. http://www.lwbooks.co.uk/journals/soundings/manifesto.html.

McDowell, L. (1997) *Capital Culture: Gender at Work in the City*. Oxford: Blackwell.

McDowell, L. (1999) *Gender, Identity and Place: Understanding Feminist Geographies*. Minneapolis: University of Minnesota Press.

McDowell, L. (2003) *Redundant Masculinities: Employment Change and White Working Class Youth*. Oxford: Blackwell.

McDowell, L. (2011) Doing work, performing work. In *The SAGE Handbook of Economic Geography*. Leyshon, A., Lee, R., McDowell, L. and Sunley, P. (eds). London: Sage, 338–349.

McDowell, L. (2015) The lives of others: Body work, the production of difference, and labor geographies. *Economic Geography*, 91(1), 1–23.

McRobbie, A. (1997) Bridging the gap: Feminism, fashion and consumption. *Feminist Review*, 55(1), 73–89.

McRobbie, A. (1998) *British Fashion Design: Rag Trade or Image Industry*. London: Routledge.

Puig de la Bellacasa, M. (2011) Matters of care in technoscience: Assembling neglected things. *Social Studies of Science*, 4(1), 85–106.

Rantisi, N. (2014) Gendering fashion, fashioning fur: On the (re)production of a gendered labor market within a craft industry in transition. *Environment and Planning D: Society and Space*, 32(2), 223–239.

Richards, P.L. (2008). Knitting the Transatlantic bond: One woman's letters to America, 1860–1910. In *Geography and Geneology: Locating Personal Pasts*. Timothy, D.J. and Guelke, J.K. (eds). Ashgate: Aldershot, 83–98.

Srnicek, N. and Williams, A. (2015) *Inventing the Future: Postcapitalism and a World Without Work*. London and New York: Verso.

Stanes, E. (2018) Young adults, consumption and material-cultural engagements with clothes. PhD Thesis, University of Wollongong, Australia.

Warren, A. (2016) Crafting masculinities: Gender, culture and emotion at work. *Gender, Place & Culture*, 32, 36–54.

Warren, A. and Gibson, C. (2014) *Surfing Places, Surfboard Makers: Craft, Creativity and Cultural Heritage in Hawai'i, California and Australia*. Honolulu: University of Hawaii Press.

7 The science and the art of making

Bartenders, distillers, barbers, and butchers

Richard E. Ocejo

A customer at a table at Death & Co., a craft cocktail bar, orders a Martini through the waitress, who brings it to Alex, tonight's bartender.

"Did he say how he wanted it?" he asks her.

She shakes her head. Alex nods to himself and starts making it. He takes out the bottles of Plymouth gin and Dolin dry vermouth, measures out two ounces of the former and one of the latter and pours them into a mixing glass, puts in the ice, and stirs it with his free hand behind his back for over a minute, glancing down at it past the end of his nose. When he finishes he strains out the drink, peels and adds a twist, and places the coupe on the waitress's tray. Since Martinis can be made in a variety of ways (with gin or vodka, with varying ratios of spirit to vermouth, shaken or stirred, up or on the rocks, with an olive or with a twist – or with a pickled onion, to make a related drink, a Gibson), I ask Alex if they make Martinis the same way every time.

"Yes, we always make it the same way when it goes out on the floor. You have to stand behind the version of the drinks that you [the bar] make, because they are the best. But an order at the bar always leads to a conversation, because there are so many ways to make a Martini and everyone has their own preference. We can't do that at the tables, unless they specify with the waitress."

***

One Thursday in the early afternoon at Tuthilltown Spirits I assist Liam as he distills. First we have to transfer the mash to the still upstairs. I take the cloth top off one of the fermenters and give it a stir with a large plastic paddle. It bubbles a little.

"It's probably still fermenting, which explains that CO_2," says Liam. "But it's been in there for a week, so it's just about done."

He measures the brix, or the suspended solids in the liquid, which is a test for sugar levels, and then we hook up a pump to the tank that runs upstairs to the still. We transfer two hundred gallons into it, which will distill down to thirty to thirty-five gallons of spirit. Liam and the distillers transfer fermented mash into the first still twice per day, once in the morning and then again around now (midday). After a first run, the spirit goes through a second, rectifying run in the other still. We're making Corn Whiskey today, and this morning's first run is going through its second distillation. It slowly drains into a thirty-gallon stainless steel drum. Liam then does what he calls an "intermediate cut." Or, he wants to see if he likes it after a gallon or so. If he does, then it goes back in the drum. It not, he'll start the tails. He takes two small glass snifters and fills them with a bit of the draining liquid. He hands one to me, and we smell and sip.

"It's a strong alcohol taste," I remark.

"I agree. That [taste] should be prominent. It also tastes like wet clothes. We'll cut the tails fairly soon, which is the difference between U.S. whiskeys and scotches, because scotches are aged longer, so they cut deeper into the tails. The tails can be rough, but aging smoothes them out. Since U.S. whiskeys are generally not aged as long, we cut the tails sooner, and then run them along with the heads through the still again."

A bit later Liam says, "See, the liquid coming out now is cloudier and milkier instead of clear as it was before. The cloudiness is caused by fatty acids and oils and is very rich. It lets you know that you are at the end of it."

***

"You guys probably don't like it when people show you photos," a client says to Miles, a barber at Freemans Sporting Club, after sitting in his chair.

Barbers get this question often. For some reason clients think barbers don't like referring to pictures, perhaps thinking they're unhelpful or a form of cheating. Not true. Barbers love seeing photos. "Pictures are the best explanation," says Ruben, since clients often cannot clearly articulate what they want. Photos of themselves are the best, because a barber can see how short and what style they want. But sometimes clients show photos of models and celebrities from magazines. These visual aids are also helpful, but come with a warning.

"Pictures work so much," says Miles. "The only time pictures don't work [is when] they want to look like that person. I've had people come in, sit down, and show me a picture and say, 'I want this.' I'm like, 'You see this photograph of this guy sitting on a beach with a beautiful woman in a nice beach chair with a drink? That's what you want. You don't want that haircut.' It's not that the haircut is bad, but I can just look at you and say that you do not want that haircut."

In these cases, barbers size up their client, put their thumb over the man's face to separate the two and better visualize how the style might work, and make an honest assessment. In this case, Miles's client has a glossy page from a magazine with picture of a fashion model. Luckily, the style can work on this client.

"OK, your hair right now is a bit too short from having been buzzed. You could totally do the side part, but your hair is still a little short to do it. But I could cut it so that by your next haircut you could do it."

***

At ten in the morning, just as Dickson's Farmstand Meats, a whole-animal butcher shop, opens for business, I walk in from getting coffee and see Aldo standing by the books, casually flipping through *Whole Beast Butchery* he grabbed from the shelf.[1]

"You ever refer to a book for your work?" I ask.

"I used to sometimes, when I started working. They were helpful, good for learning. There was one old one I used, a simple one, I think we have it."

He can't find it in the mini-library, and he can't remember the name. Aldo keeps flipping through the book, pointing out what he thinks is wrong or what he could do better. He opens to a photo of a Frenched rack of lamb, and points to the meat still on the ends of the bones.[2]

"You see that? Now, look."

He brings me over to the display.

"Those are beef [bone-in rib chops], but you can see how clean it is. We do it good."

***

These episodes are from a larger research project I conducted on occupations that have transformed from typically and traditionally low-status, undesired, "dirty" jobs, to higher-status, "cool" jobs that involve innovative work practices and cultural taste-making.[3] As a result, people with other options in the labor market, such as college graduates, people with full-time jobs in other industries, and people high in cultural capital, have sought them out as meaningful careers. I studied cocktail bartenders, craft distillers, upscale men's barbers, and whole-animal butchers and butcher shop workers – all high-end, niche versions of these very common occupational categories. Important to this transformation has been the enhancement and elevation of the techniques they use to make their products and provide their services within the context of their workplaces to a level often described as "art." These upscale businesses revolve around these techniques and deliberately put them in display for consumers.

Workers in all kinds of occupations have an "occupational aesthetic," or an idea of right and wrong in how their job should be done, what the products and services should look like, and what the results should be.[4] It's what makes every worker in any occupation an artist, at some level. The workers in each of these occupations are no different. Their sense of right and wrong is deeply rooted in the cultural repertoires of their niche jobs. A sense of craft and craftsmanship is central to the cultural repertoires these workers enact. This sense is a conscious focus on the technical aspects of their work, bundled with the cultural knowledge they learn and communicate. As enacted in their techniques, it also holds a central place in the social world of their workplaces. They perform their craft publicly, for the purposes of transparency and to spark conversation.[5] But they also all believe in doing their work well for its own sake, such as when they stress over details no consumer would ever notice.[6]

What does this sense of craft look like and how do these workers enact it? Building on these opening vignettes, in this chapter I show how people in these transformed jobs think about the technical work they do. I begin by showing how these workers all identify aspects of their work that are beyond their control, or where they must respect nature and follow specific procedures (science). Next I show the aspects that demand their input (art). The "art of" what they do, however, is based on a set of templates created by their occupational communities. Learning the limits of each – when, where, and how science and nature end and art and creativity begin – is fundamental to a successful, confident performance. I conclude with a brief discussion of how social contexts constrain the impact of skill in these and other occupations.

Respecting nature

In the early afternoon on delivery day, JM enters the cutting area from having lunch in the back to break down more cows. Giancarlo and Brian had already brought two more sides out from the walk-in to hang. JM picks up his knife and runs it along the sharpening steel. As he turns around his eyes widen when he sees one of the carcasses before him.

"Oh, look at that," he says. "That's Prime."

"How can you tell?" I ask.

"Look at the fat."

Later on in the day I observe Lena breaking down a cow shoulder on the table.

"It's old," she says without looking up from her work.

"You mean it was in the walk-in for a while?" I ask.

"No, the cow was old."

"You can tell?"

"Yeah. The meat is stuck on the bone."

All of these workers confront science and nature in their work, which presents them with both challenges and clues about how to proceed. Experienced butchers can answer a lot of questions about an animal from looking at it hanging in front of them, split into large sections. What did it mostly eat when it was alive? Was it active or more docile? Was it panicked when it was slaughtered? The insides give answers. Since grass-fed cows are less fatty than grain-fed ones, it's rare to get Prime-grade cuts from them, since meat grades are largely based on fat content (or marbling). The best level whole-animal butcher shops usually hope for is Choice-grade cuts, one notch below Prime. JM was surprised because he doesn't see such a fatty grass-fed cow very often. He certainly used to in previous butcher jobs.

At some point in their work lives, some aspect of the process is out of these workers' hands. Nature produces both consistencies and inconsistencies. Butchers are able to learn how to break down animals because they are all the same in their anatomy (i.e., muscle groups), but they also vary as organic products. They must work on efficient and consistent butchering to make the cuts they sell as uniform as possible. But consistency in the final product is an elusive goal. More importantly, butchers at craft shops attribute its elusiveness to their meats' provenance (grass-fed, or grass-fed, grain-finished), and thereby to their meat philosophy. In other words, their occupational aesthetic incorporates their sense of "good" meat into their idea of "good" butchery technique.

"I think the trick with the local small farm kind of model is that there is so much variation in the meat," explains Jeff. "Some weeks the meat comes in and it's beautifully marbled, and some weeks it comes in and it's a bit floppy. But that doesn't necessarily need to matter if what you care about is that it's local and it's sustainable and it comes from a small family farm. There is a certain allowance you leave for the variation there. There wasn't that, like, consistency that they get from supermarkets or something. When I first started as an intern I was actually surprised that more customers didn't come in complaining that there was such variation in 'quality.'"

Since the shop sells artisanal, not mass, products, which is how quality meat "should" be made, customers "should" therefore accept degrees of variation in appearance and gustatory quality. But the situation presents a paradox: butchers aim for preparing products that are consistent in the quality of their presentation and taste, while recognizing that they work with natural products that will often vary. Animals that are pasture-raised on small farms (as opposed to feedlots, where their growth and eating habits can be more efficiently monitored and controlled) will occasionally be different in appearance and taste. Workers at whole animal butcher shops think that many meat consumers equate Prime with quality, because it is what they have been taught to think by the mainstream meat industry. A lack of customer complaints about the variation in the meat is a testament to the shop's ethos and ability to successfully educate the public.

As this example shows, the science and nature aspect of their work, or the elements they cannot control, presents these workers with a constant tension as manual laborers working for a business: while they will always be beholden to degrees of inconsistency, which is one of the foundations of their businesses, being wildly inconsistent puts their businesses at risk, since modern consumers have come to expect a large degree of consistency in the products they buy. Butchers can't control what nature does to animals. They can only choose their farmers wisely (and their animals, if they wish to visit the farms themselves) and focus on their butchery techniques. "On the fresh meat side I would say 80% [of quality is] farmer's quality and 20% [is the] butcher's ability to cut it," says Jake. "Not that you can't destroy good meat by being a bad butcher, but I think what makes the product special is mostly the quality and how it's raised."

This paradox, tension, and deference to something beyond their control exist for each of these occupations, in different ways. Like other bartenders, cocktail bartenders use a lot of alcoholic products that companies make in large quantities. They avoid some of the big-name brands, such as Grey

Goose, Bacardi, Jose Cuervo, and Bombay. But cocktail bartenders can reasonably expect most of their spirits and liqueurs to be consistent in their taste. Like the meat industry, the spirits industry has techniques for ensuring a consistent flavor in its products. An example is blending, or combining multiple batches of a spirit – from barrels or other containers – before bottling. Like when making a soup, blending together multiple ingredients can smooth out any imperfections. Many larger spirits companies employ "master blenders" along with "master distillers," because their consistent flavor comes as much from the right blending combination as from the distillation. As a result, cocktail bartenders feel the pressure to not ruin what is already "naturally" good.

"We have a lot of respect for what goes into these bottles," says Joaquin. "When we meet master distillers, it is an *amazingly* humbling moment to meet the guys that actually do this. It's great! They have incredible palates and a great skill set, and all we can do is tell them, 'We're trying to not get in the way of what your work is. We're not trying to mask your life's work with a bunch of unnecessary liquors. We're trying to round out those flavors that are already existing in the bottle. We're trying to bring them out, we're trying to showcase them well,' and the way you do that is by choosing the spirits for your cocktails wisely and then executing them precisely every time. Pretty much every time I get, 'Wow, you're so good!' [from a customer I say] 'You know, when you're mixing with these mixers that were juiced that day, when you're pouring these spirits, I don't need that much credit for doing this. They're doing all the heavy lifting for me. I'm only providing texture and temperature. These guys are giving you all the flavor.' That's the beauty of it."

In this sense the distiller becomes nature, and nature is perfect as is. Cocktail bartenders cannot control, however, how a small batch of a spirit is going to taste from one bottle to the next, or the quality of the citrus they use (they can only squeeze it right before mixing it). Smaller craft distillers often sell products on the market fully aware that batches vary, because they have been experimenting (different recipes, different barrels, no blending) or still haven't settled on their recipe. For the cocktails they use craft spirits in, cocktail bartenders can only sample the bottle beforehand, follow recipes, and adjust their modifiers as needed to achieve the flavor they want. These products represent the inconsistency beyond their control in their work lives.

For upscale men's barbers (and for all barbers), the science factor in their work is human anatomy. People's hair (straight, curly, wavy, oily, thin, receding), heads (large, small, narrow, wide, bumpy), and faces (round, oblong, big ears, small foreheads) are what they are. Barbers have no choice about who their clients are or what nature gave them. Hair tells them what they can and cannot do to it. Experienced barbers learn how to listen, follow, and respect what hair wants; there is no sense fighting it. The difficulty in dealing with science for barbers is when social and cultural factors intervene, namely when clients ask them to go against nature, to defy what their hair genetically wants to do (e.g. go in a certain direction due to their whorl) or be (e.g. straight, thin). People's biological differences always factor into their work, for better or worse, and barbers often discuss them openly.

On a Wednesday evening a suited Asian man sits in Van's chair for an after-work shave. Husky with a large neck, the man doesn't have a thick beard or a pronounced jawline. From the start it is clear he doesn't want to chat. After answering Van's questions ("Have you ever had a straight razor shave before?" "Do you go against the grain?"), he immediately closes his eyes and relaxes. Van starts the process and tries to take his time. Before I know it the shave is over and the man has thanked and tipped Van and left.

"That was easy," I say.

"Super easy," says Van. "Big fat guy. And he had a mix of hair. He was Filipino, or something. Japanese guys are the hardest to shave, because their hair is so thick and straight. Straight hair is like a circle, curly/wavy hair is like an oval. You pull out as much straight hair as you shave. How long did that take me? Ten minutes?"

Van's client was a barber's dream. Regardless of his actual ethnicity, his hair was a mix of straight and curly, and his beard wasn't thick. Certain ethnic and racial groups have tendencies in their genotype, which shape patterns (and stereotypes) about their appearance. In showing preferences for people of certain appearances, barbers can come off as prejudiced. But they are really commenting on what people's genes have determined they look like. And his body type – specifically his rounded jawline – meant no edges. He had little visual separation on the skin between his face and neck, which meant a smooth, continuous surface for Van to shave (like having a long cheek). Slightly curved surfaces are much easier for shaving. These factors combined to make this shave quick and easy, no matter how much pomp and circumstance Van added.

Using the senses, harnessing time

Given the role of nature – the form of the meat, the grain, the spirit, and hair – an important aspect of these workers' work is learning to control and harness it, specifically where and when in the process and how to do so. They place a strong emphasis on training their bodies' senses to know when and how, and on their potential to use their senses and abilities to manipulate time. Time for these workers is both a constraint and a luxury. It structures the work they do.[7] Each occupation has moments when workers feel pressure to work quickly. It could be during a shift, such as when the post-work crowd walks into a cocktail bar, a time of the week, such as Fridays and Saturdays at barbershops, or a time of the year, such as during the holidays at butcher shops.[8] But owners of these businesses deliberately design their method of operation to give their workers more time to do what they need to do to make a quality product and provide a quality service. The aim is not slowness per se, although they do their jobs more slowly than other versions of these occupations in the sense that it takes them longer to accomplish their main objective (make a drink, make a spirit, cut hair, cut meat). The goal is to do their jobs "right."

Each occupation requires workers to have well-honed senses. Cocktail bartending demands the use of all five of them. Taste is perhaps the most obvious. Like chefs, cocktail bartenders aim to sharpen their palates to identify the flavor profiles in ingredients, for the sake of discovering combinations for new drinks and for explaining what's in a cocktail and why it tastes as it does to customers. Tasting is what they mainly focus on in their educational programs. And of course, smell is integral to taste. Cocktail bartenders develop their palates and taste memories over time. They hone their senses of smell and taste by constantly sampling products.

Less obvious in their work is the role of hearing, touch, and sight (beyond the obvious ability to see what they are doing). The sound (and sight) of a bartender vigorously shaking a cocktail in tins is a main attraction in cocktail bars. When the sound fills the air, customers regularly stop their conversations and look up at the expressive, and somewhat violent, display behind the bar. While bartenders adopt their own shaking style (some change theirs over time due to injuries and soreness in their elbows and shoulders), the aesthetic has a function. They are mixing the ingredients well, breaking down the ice to add water (a key ingredient) and tiny ice crystals for the cocktail, and getting it cold. They know when a cocktail is "done" – diluted and cold – in the shaker by both listening for the large ice breaking up into smaller pieces (the sound changes from heavy thuds to light rattles) and feeling the metal get colder and colder, sometimes turning frosty, in their hands. (Unless they're making a drink with egg whites, cream, or a thick syrup, each of which requires harder shaking for longer durations, they know a cocktail will be well-mixed at the end of a standard shake.) Brian, who is known for having a hard and theatrical shake, explains his shaking:

> "When I think about my shake that's when I screw up. That's when I knock the shaker over or the jigger hits the side of the glass. It's an internal mechanism. You can shake it too much, even with the block ice, and you can dilute it. You have that internal clock. I used to count how many times I shake, but now that I've done it so much it's a sound, like I just know when the drink's done. There's also like a dance, [and] sometimes I can tell when my shake's off. Like when I'm shaking too short I haven't done it enough but my arms are kind of tired and they're tightening up, and I just stop early."

And once they've shaken or stirred a drink or already poured it out into a glass, cocktail bartenders regularly taste it (with a plastic straw) as a form of quality control. In this way they can tell if they made it right, that is, well-balanced: measured, mixed, and diluted correctly, and cold. Sight, meanwhile, sometimes plays an important role in a new cocktail's aesthetics. While sometimes it can't be helped because of the ingredients they use, cocktail bartenders want their drinks to look beautiful. They sometimes combine ingredients to make vivid colors, with garnishes that provide attractive accents and contrasts.

As with cocktail bartenders, taste and smell are the most important senses for craft distillers. They taste at multiple steps in the distilling process, such as while proofing, when a spirit is aging (to determine if it is "ready"), and during distillation runs when cutting the heads, hearts, and tails. Large distilleries often use computers to make these cuts automatically, such as by measuring the temperature in the still, and ensure consistency. But craft distilleries do so manually, by simply having someone decide when to do it. Workers have their own way of gauging when to make cuts, such as by seeing how much volume has been distilled and the color of the distillate coming off the still, and, primarily, by tasting. Craft distillers taste constantly, their own products and other brands. I ask Liam to explain how he determines when to make the cuts:

> "You can taste that foul, almost fuel-type thing, nail polish remover [in the heads], and then also like the texture of it, kind of dry and chalky. You're getting to that spot of the cut and you taste increment by increment, very small, you'll notice that it's a pretty sharp like chalky sour, then sweet. Sweet and chalky, then [the chalky] disappears and it's sweet and you know that's your product. The tails cuts sort of taste like a damp, wet rag, or again you get some methanol at the end, it comes back so you can kind of taste that. It depends on what you're [distilling] too. If it's white [corn] whiskey and it's not going into a barrel, it's going to go right to the bottles, you have to be a little more conservative about those flavors. Whereas if you're going to put it in a barrel for six months to a year to two years to six years you really have a lot more leeway with it. Different alcohols are oxidized [differently] as time goes on."

Upscale men's barbers and whole animal butchers both rely primarily on sight and touch, most especially the relationship between the two. Both jobs require advanced hand–eye coordination to do the work well and avoid injuries (they are, after all, both working with blades). They train themselves to learn how to use their tools confidently, and to get a "feel" for the raw materials they use (hair and meat). Sometimes subtle changes to these elements can have consequences. Over time the scissors, clippers, and razors become extensions of the barber's arm. Eyes, hands, and tools coordinate to cut hair and achieve style. When one is off, such as a different pair of scissors, sometimes the process gets compromised.

Barbers deal with time in the long-term and short-term. Long-term they build relationships with clients and learn about their hair and their lives. Over time – perhaps a couple of months, perhaps a couple of years – barbers figure out what style will work for their clients, while the clients come to trust the barbers with their hair. Also, some styles require multiple visits to achieve, which means a single haircut is part of a long-term plan, a step in a series. In the short-term, the haircut itself is a process. Because of the amount of time they allot for a haircut, upscale men's barbers can use techniques that barbers at other shops do not (and sometimes cannot) use. These latter barbers have to work quickly and are not aiming for a unique style.

Butchers are very similar to barbers in that learning to be precise with their tools, for quality work and safety, is of utmost importance. They aspire to cut efficiently, but not recklessly. As with barbers, doing so requires hand–eye coordination and exact body position, as well as a feel for the animal.

Brian has been working at the shop for a couple of months. Originally from Michigan, he moved back to Detroit after attending college in Oregon and living in Portland for a few years. He started a sausage-making business with a friend, and he became very familiar with pork and breaking down pigs. But he is only just learning beef, and today Giancarlo is showing him how to break down the cow shoulder.

"This is the elbow," he says to Brian while pointing at his own elbow. "You go around the bone and separate the meat. This is shoulder tender, the brisket, the clod, and the platanillo. *See* the cuts coming off."

Brian nods along and slowly follows the seams with his knife.

Getting a feel for the animal for butchers in part means learning where the seams are for the different cuts they are after. Butchers can often remove, or "seam out," sections of animals, especially cows, in a variety of ways. The seams can represent different muscles entirely or different sections of the same primal cut. Shops and head butchers determine what the retail cuts will be, and butchers are interested in the decisions other shops and butchers make.

Later in the day Giancarlo starts working on the tri tip, a section from the hindquarter that includes the tri tip, top sirloin, sirloin fillet, and culotte cuts. It's a tricky section to remove because the butcher is both pushing the carcass away from the body with the knife while pulling it towards the body with the hook, and it's very heavy. As he works on removing it from the carcass, JM notices his knife work.

"You're killing the cow again, you fucking asshole."

Giancarlo gets it off and plops it down on the table.

"Argh!" utters JM.

Dave, the chef, has been vacuum sealing some bags and watching with a curious look on his face.

"Show me what's wrong," he asks JM.

JM brings him over to the hanging carcass.

"You see, the line should be here to get the flank [points to the hanging carcass]. But it's OK, I don't want a lot of flank."

Giancarlo then starts taking off more of the suet that is still on the animal in the trip-tip area. JM shakes his head as he watches him slice it off.

"What's wrong with that?" I ask JM.

"It should have come off before. Because then you're pulling. Now you have to break it apart."

In other words, Giancarlo missed his opportunity to remove the suet efficiently when he was removing the tri-tip with his hook. Now he must slice it off piece-by-piece, which is time-consuming. He also cut too much into the flank cut when he removed the tri-tip, because his seam was off. He didn't visualize it as well as he should have. As a result, they'll have a small flank to sell. These are both skills butchers learn over time, through practice.

Being creative: templates and improvisation

One year at Tales of the Cocktail, an annual event in New Orleans for the global cocktail community, I attend a panel called "How to Bang Out Drinks Like a Maniac," run by Philip Duff and Dushan Zaric, both longtime bartenders and cocktail bar owners. They designed the panel as a counter to the panels and overall discussion in the industry that elevates bartenders to the level of artists. Drinks must be of a high quality, they contend, but cocktail bartenders must also be honest about themselves. Their business-oriented talk and slideshow focus on achieving efficiency without sacrificing quality. Halfway through, Philip chides bartenders who spend too much time tasting and refining a cocktail, trying to achieve some level of personal perfection, before they serve it.

"We are not tortured artists. I'm going to drag up another guy from the past who said it better, I think, than anyone before or since."

He then changes the slide to a quote from Patrick Gavin Duffy, a cocktail bartender from the early twentieth century:

> Bartending is an old and honorable trade. It is not a profession and I have no sympathy with those who try to make it anything but what it was. The idea of calling a bartender a professor or a mixologist is nonsense.[9]

"It is an old, and honorable trade, no more, no less," continues Philip. "Like making hats, or being a plumber. You can see Patrick getting a little dig at somebody there, someone calling a bartender a professor or a mixologist. But he's right: it's a trade. Like a carpenter measures twice and cuts once, and then he gets on with his life."

None of these workers refer to themselves as artists. Most self-identify with their job's basic label ("I'm a bartender"), not its superlatives ("mixologist"), and say it's their career, profession, or occupation, not their calling. Still, they all readily recognize the art, craft, and creativity behind their work, and strive to do their jobs well regardless of whether anyone will recognize it. As part of their cultural repertoires, this sense of craft and craftsmanship and the creative potential of their jobs is central to the meaning they derive from their work and to their desire to pursue their jobs as careers. They all focus on the "art of" what they do rather than call themselves artists.

The art of doing each of these jobs means learning how to follow, bend, and sometimes break a set of rules that their fellow workers follow and enforce. Each occupation requires its workers to work with a set of templates. These workers, then, are not dissimilar to jazz musicians, who work to learn patterns common to many songs, and once they do are able to play songs they have never played before and even add new elements to them. As Faulkner and Becker examine, "The songs of the jazz repertoire…are, for the most part, formulaic, elaborate variations on a small number of templates. Someone who knows the basic forms can play thousands of songs in this great reservoir without much work" (2009: 24).

> To some extent every performance involves elements of improvisation, although its degree varies according to period and place, and to some extent every improvisation rests on a series of conventions and implicit rules…[J]azz players routinely play versions of songs they already know of whose form they can guess at, substituting melodies composed on the spot for the original, but always keeping in mind that those melodies ought to sound good against the (more or less) original harmonies of the song, which the other players will be (more or less) expecting to be the foundation of what they play together. Jazz improvisation, then (more or less), combines spontaneity and conformity to some sort of already given format.
>
> (2009: 27–8)

When they first start working, these workers train themselves to learn the templates of their jobs. Learning means knowing the technical skills in both their minds (what to do) and bodies (how to do it). The former entails mentally grasping and memorizing the bases of their actions, while the latter refers to developing a "second nature," or when they can easily enact the technical aspect of the cultural repertoire. Once they learn the template, mind and body, they can then manipulate it. They can improvise.

When making cocktails with existing recipes or when creating new ones, cocktail bartenders aim to achieve balance. When they succeed, no single ingredient overwhelms the others, or an ingredient has been manipulated (tempered or enhanced) to create a new flavor. The drink is not too boozy, sweet, bitter, or diluted. Several templates exist for making balanced cocktails.

"To get balance," says Toby, "there's the triptych, which is strong [boozy], sweet, and tart, and you want all three. If a drink has too much of any one of those things it's not going to be balanced. And so if you have two ounces of booze, ¾ of an ounce of fresh-squeezed lime juice or lemon juice, and ¾ of an ounce of simple syrup so it's one to one, you're going to have a perfectly balanced drink. It just works that way. If you have a Daiquiri, a Gimlet, those all have that template. And so once you start jiggling that triptych, say you're going to make a Margarita, you know that Cointreau isn't as sweet as simple syrup, so you know you have to bump that up or bump the other one down.

And so it's all about these little, tiny movements, of the mixture of it. And once you get that in your head and you just know it, then you can make up drinks."

"You can substitute," I say.

"Exactly. And just by little, tiny bits – one-sixteenths of an ounce sometimes – [of something] that you [add or subtract] or you add a splash of this or a splash of that, but you're still in that template. And you've created a style. I could look at some recipes and almost tell you who came up with the recipes because I know everybody has their own little quirks and their own little things they like."

Today's cocktail bartenders have many recipes memorized, but knowing templates, or understanding the conventions behind why certain drinks "work," is far more important. Improvising for them refers to swapping and adding ingredients and adjusting ratios. Creativity stems from improvising within templates and developing a unique style.

"Everything is a reference to something else," says Joaquin. "Everything. Everything is basically a riff on a handful of some classics. Everything you make. There's nothing new under the sun. Where people get creative is with what you're using as your ingredients, but for the most part you have the same basic formulas. The real art of it is fine tuning it into something that's really special and stands on its own and can count as its own creation, its own cocktail, regardless of what the inspiration for it was. When I made the Latin Quarter I knew I was making a Ron Zacapa Sazerac with a dash of chocolate bitters added to it. That's exactly what I was making. I knew that. But it works on a different level than a Sazerac where it's not the same drink. It's not as spicy. But does it work as a cocktail? Absolutely. Did I reinvent the wheel? Hell no. But I made a good drink. There's only so original you can get."

Like cocktail bartenders, craft distillers work with ingredients and recipes. But balance is more subjective for them. Craft distillers aim to adjust ingredients and recipes to achieve a flavor they're after. There are many precedents and conventions in distilling, but there are too many factors involved in the craft distilling process to ensure precise replication.

"I think the thing that makes it sort of artful as opposed to straight science-y is the pure engineering approach doesn't work for whiskey," says Nicole. "There is such an art to just getting the fuel for the process. What makes a good whiskey? Getting the feel over what your process should look like. You can use a recipe but you really have to know how it is supposed to feel. Kind of artful."

With its chemical processes and complex machinery, craft distillers blend art and science the most of these four occupations, and emphasize both as integral to their work. The workday at craft distilleries is an interplay of sensing and measuring, tasting and operating. Creativity comes out of creating recipes, constructing a process, and making decisions based on tasting.

"I think that science is a foundation," Bill says. "The art and the tasting is crucial. When it comes down to it, who cares what's going on with the alpha amylase as long as the whiskey tastes good? But, the flipside of that is that having a really good foundation and knowing what's going on will allow you to really make creative decisions that make sense. When everything's going right, then it's perfect. Tasting, you're like, 'All right, this tastes good.' But, if something goes wrong, your sense of things and your gut instinct doesn't help you as much. Having that understanding underneath, behind the curtains, that allows you to troubleshoot, and it allows you to gain efficiencies. This is a business, and so the art and the craft of it allows you to have a very good product. Once you have a very good product and you've been selling it, all of a sudden you need to get that efficiency up because you're a business. If that understanding allows you to boost production in certain ways, if all of a sudden you can, because of your knowledge about sugar to starch conversions, you're getting an extra 2% on your fermentation, that's a really meaningful jump. For those reasons the scientific understanding is really important, especially if you're going to start putting it at a meaningful volume. You have to know what's going on."

Multiple factors can impact the final product, such as the quality of the ingredients being distilled, the fermentation process (e.g., open- versus closed-top fermenters, outside temperature), the make-up of the barrels in which the product is aging, and, as Bill states, distillers' tasting abilities. Creativity for craft distillers stems from the decisions they can control during various stages in the distilling process, even if they lead to uncontrollable outcomes. Like cocktail bartenders, they experiment with ingredients and recipes, sometimes following existing templates. These include choosing different varietals of base products (corn, rye, barley, wheat) or botanicals (juniper, cardamom pods, fennel seeds) in different combinations and ratios. They decide on fermentation technique, the type of still they will use, the number of times they will distill their products, and when to make the cuts. And they choose the proof level, the water they use (to cook the mash and adjust the alcohol percentage), the size and type (i.e., char level) of barrels to use, how long to age the spirit, and whether and how to blend the barrels. Each choice represents a different way for a distiller to create a unique product, but most also add an amount of uncertainty into its quality.

Unlike cocktail bartenders and craft distillers, barbers and butchers are not working with ingredients and recipes. Adding material items is not part of their work process. (Exceptions are when barbers use pomades at the end of a haircut to preserve a style's shape, and when butchers tie cuts of meat.) Improvising on their templates is not about substituting or making adjustments to ingredients. They base their templates on what they know about and their experience with the material products they work with, specifically people's hair and heads and whole animals. Over time they build their knowledge of these templates, and hone their abilities to work with them and improvise.

I occasionally photographed the barbers working, and then met with them afterward to discuss their technique.[10] One time Ruben was very busy with clients and we weren't able to look at and chat about the photos until a week later. Knowing he had personally done scores of haircuts and seen even more since then, I jokingly ask him if he remembers this one as we sit down to look at the pictures on my laptop outside the shop.

"Of course I do," he replies, while looking at the first photo.

"You do?" I ask, surprised.

"Yeah yeah. I mean, it's a typical haircut."

Like the recipes of cocktail bartenders and craft distillers, barbers have templates for the styles they create, which are based on the work process, the client's physical makeup, and what the client wants. The first photo I took of Ruben during this haircut simply showed him using his clippers on a section of his client's hair above the right ear that he was lifting with his comb. Based on what it showed, the fact that it was the first in a series, and what he saw was the client's hairstyle at the start (as viewed from behind), Ruben, an experienced barber, "remembered" it and could then explain the rest. In other words, he could mentally put himself in a position he has been in countless times before.

"The first thing you do is you try to figure out where the part is," Ruben says about this type of haircut. "The part usually is very simple. You look at the cowlick, and then from the cowlick that's the direction [the hair goes in]. You narrow down where the part is. Once you've figured that out, you obviously find out what they want, and what I did on this guy's hair, it was a three-and-a-half [clipper guard] on the side, you kind of draw an invisible line over here [on the side of the head], and you always have to keep the line below the cowlick area, because once you raise this up it becomes frizzy, so you're trying to avoid that as much as possible. So you do a nice gradual line all around. It starts by where the temple is, and it finishes on the other side by where the [other] temple is."

Here Ruben follows the work process by finding the part and learning the direction the hair goes in. Hair directionality is natural and immutable for everyone. Where it goes and what happens to it in its course (e.g. length, texture, bald spots) varies from head to head. The photo doesn't show it, but before finding the part Ruben had to figure out the stylistic direction. Barbers have to "find out what [clients] want" or help them figure out what they want or what "works" for them. With this information they can choose and follow a style template (or simply follow the one the client tells them to do). The style template varies by client, then, because people's hair, heads, and faces are different, and, crucially, people have different ideas of what they want. These factors sometimes conflict, such as when a client with a certain hair type has an incompatible request. Ruben sees from the first photo that he's doing a "typical haircut": short on the sides and back, longer on the top. The guard size he starts with (three-and-a-half) tells him more about the style he is working to achieve, because the guard determines the length of the hair.

The general style at Freemans is for more "natural" looks, which includes gradual fades rather than sharp lines. For a typical haircut, Ruben knows to start the guideline (literally a line he will eventually blend into the rest of the hair) by the temple with the largest guard size, and then use smaller and smaller guards to fade the hair down the back of the head and into the neck, disappearing into the skin like a song fading out. Since he started with a three-and-a-half, Ruben immediately has an idea of the style template he's following. A three-and-a-half won't create a stark contrast between the hair on the top of the head and the hair on the sides and back. A one guard all around, for instance, would, which is a much different style (known as an undercut). This client clearly didn't want this more striking style. He wanted a typical haircut.

Creativity for barbers comes from several related sources: regularly working with people who have their own unique combination of hair, head, and face as well as their own sense of style, getting to improvise within style templates based on these differences and by using an array of techniques, and developing their own style.

"Each client is different," says Eric of the Blind Barber, "so you can't give the same haircut for every person. You have to change up per client and that is where the creativity comes in. And everyone's hair is different, it grows in different ways, different patterns. [You're] working with what they have. And you know, some people are a little bit more out there and let you go in different directions and are more trusting and do not want a simple haircut, they want something a little bit more stylish or a little bit more avant-garde. Then you really get into being creative."

Barbers first learn the basic and more specialized techniques of cutting hair, in which they eventually become confident. As they do, they gradually develop their own style for how to improvise within the style templates. Personal style is an added level of creativity for barbers. Bret, who handles hiring at Freemans' shop in the West Village, explains this idea by comparing different barbers:

> "[Some new barbers] who come [to work] here have a little more limited training and want to grasp some of the more advanced, creative techniques. Some of the guys all learn how to cut really short hair – typical barbering. I wear long hair. They couldn't cut my hair. I mean, they could do it, but they wouldn't be practiced or have fun doing it, because they don't know where to start and where to end. Just practicing and doing a few simple techniques and tightening up the skills that they do have, just make sure it's nice and crisp. I interviewed someone Monday, someone came in and did a few haircuts. The haircut overall was good, the shape of it and the design of it. But you got into the details, the finishing points of it – the neckline and all that – that's what sets you apart. The haircut should be technically good, but every part of the haircut should be really good. Some people can do really, really clean around the ears, nice square neckline, but the interior of the haircut is a mess. But it looks sharp from the outside to the common eye. So that's where you want it to look really clean. That's the idea to have it really clean but have the shape that grows out and looks really good."

Barbers at Freemans can spot someone on the street who got his haircut at their shop, because they share the technical aspects of what they do (e.g. natural fades) in common. Regular clients may even be able to tell, also. But the barbers can also often tell which barber cut someone's hair, because of the style details on top of the technical details. The former are standard, but the latter sometimes take time to emerge, such as when a barber is learning about a client and his hair and building trust over multiple haircuts. Examples of these details include the use of a straight razor to enhance or create a part (Ruben) or the feather razor to create the look of messiness (Miles). Finer details are less about following style templates and more about making personal decisions.

"I was at a restaurant the other day," says Anthony, "and I'm always looking at people's hair, and this guy came in and I could tell it was a good haircut. Like, technically, it was very good. But I could see where the barber made decisions that I wouldn't make. That's fine, because that's his style. I would do it differently."

A country's meat culture and the choices of the shop often shape the templates butchers follow. The anatomy is obviously the same around the world. ("It's a cow," says Aldo, "it's not going to have like a fifth leg or two heads – 'This head is not for eating, this head is for eating.'") But different cultures prefer different parts of animals, which influences how slaughterhouse workers and butchers break them down (e.g. keep more pork belly on the ribs or have more of it for bacon; leave more meat on the shoulder or more for the ribs). Butcher shops around the world have different primal cuts, sub-primal cuts, and retail cuts, parts they throw away, and parts they use (and different uses for the parts they use). An American butcher would have to get oriented in the Italian, Spanish, or Mexican style, and vice versa.

The animal itself, then, presents butchers with one basic template (cows, pigs, and lambs are different from each other), but culture (of a country or region) and the preferences of a shop present them others, hence the interest the butchers at Dickson's had with how the people at the Meat Hook break down a cow shoulder. The creativity, or the art of butchery, comes in how clean one cuts the meat (which includes seaming it out and trimming and presenting it). Jeff explains:

> "The wrong knife pass, you know, too much sawing motion, taking off too much fat. The meat could quite literally look butchered, which it shouldn't. I've definitely been to some butcher shops and some other farmers' market stands and the meat looks like it's just mangled and butchered, literally, just horrible. You could have a subpar piece of meat with regard to quality and a good butcher can spruce it up to a certain degree."

Creativity also comes from what one can discover on the animal. Once they learn these templates, butchers can then sometimes figure out how to seam out new retail cuts from primals and sub-primals.

"Where does the flatiron come from again?" I ask Lena one day. With so many unusual cuts to learn, I often asked a question like this one to a butcher during a lull.

"It comes off the top blade where it connects with the ribs and rib eyes," she says. "You want to see a cut that doesn't have a name and that no one sells?"

Lena leads me into the walk-in and takes down a dry-aging rib section, from which the butchers will saw the popular and pricey retail cuts of rib eyes and strips. She turns it lengthwise, to show where it was removed from the shoulder section (the other end would have been removed from the loin section).

"So here is the end of the palomilla [points to the bottom part] and flatiron [top part, just beneath the fat cap]. You could easily see where these cuts end on the rib section. We cut it off for grind. But this is the best meat. You could eat it raw it's so good. But it's only a half-pound steak, and there are only two of them on every cow."

"Why not cut more into the shoulder blade?"

"Because everyone cuts at the fifth bone. That's the American style. There's more money in the rib section. I'd like to name it. It's really hard coming up with a name. I have to look at the Latin names for it."

Americans love ribs and bacon, which explains why American slaughter-houses and butchers break down cows and pigs as they do. But by thinking about how to break down a cow differently, Lena discovered a cut she would like to sell, if she had the chance. Since she works at Dickson's, she cuts meat the Dickson's way, and can only make suggestions to the head butcher and owner. In finding hidden cuts, butchers learn to separate meat from the animal in a new way, with a different flavor profile and cooking procedure. Their creativity comes from learning how to manipulate the templates they have learned to use.

Conclusion: the inequality of making

From food to furniture, there has been a groundswell of "crafty" making in the United States and other countries in recent years. Forums like fairs and Etsy, the online marketplace, have helped grow these activities. I have documented the brick and mortar businesses that are based on many of the central principles of this movement, such as "authentic" products made by hand.

It is fascinating that manual labor jobs like the ones in this chapter have become makers and influencers of taste, and come to be regarded as artists. But they only get such attention and praise when they perform their work tasks in specific settings. These tasks and their labels – craft-based, hand-made, artisanal – in unique settings of production have become highly valued in today's economy. But they do not extend to all workers within these industries. In this sense, these are similar to critiques about the "creative class": it omits a wide variety of creative activities, such as those that members of the working-class engage in for survival.[11] We often think of inequality in work in terms of access to jobs, pay, work arrangements, and conditions in the workplace. These jobs present an additional dimension to work inequality: the social contexts within which work takes place. Without the retro cocktail bar, the craft distillery, the classic barbershop, or the neighborhood-styled butcher shop, these workers cease to be seen as artists.

Notes

1 The full title of the book is *Whole Beast Butchery: The Complete Visual Guide to Beef, Lamb, and Pork*, by Ryan Farr, a San Francisco-based chef, butcher, and entrepreneur. It was published in 2011.
2 Frenching a rack of lamb means removing the meat, fat, and tissue from the ends of the bones, giving them a clean appearance.
3 See Ocejo (2017)
4 Fine (1992) refers to an occupational aesthetic as a "sense of superior production" (1268) that workers have for the practices they engage in.
5 I will also show in the next chapter how important public validation is for their professional identity.
6 This point of doing something well for its own sake is the central aspect of Sennett's (2008) definition of craftsmanship.
7 Fine (1995) discusses this idea in relation to chefs.
8 As manufacturing businesses, craft distillers are less influenced by consumer market forces. They try to operate the distilling process continuously throughout the year. Given their size, however, there are situations when they are forced to speed up production, such as if they don't receive a delivery for an ingredient in time to begin their distillation and then have to catch up.
9 This quote is from his book *Official Mixers' Manual* (1934).
10 This data collection method is called "photo elicitation." The aim is to get participants to see themselves doing what they do and to see their contexts from different perspectives. See Harper (1987, 2002).
11 See Wilson and Keil (2008) for one such critique.

References

Faulkner, R.R. and Becker, H.S. (2009) *"Do You Know...?" The Jazz Repertoire in Action*. Chicago: University of Chicago Press.

Fine, G.A. (1992) The culture of production: Aesthetic choices and constraints in culinary work. *American Journal of Sociology*, 97(5), 1268–1294.

Fine, G.A. (1995) *Kitchens: The Culture of Restaurant Work*. Berkeley, CA: University of California Press.

Harper, D. (1987) *Working Knowledge: Skill and Community in a Small Shop*. Chicago: University of Chicago Press.

Harper, D. (2002) Talking about pictures: A case for photo elicitation. *Visual Studies*, 17(1), 13–26.

Ocejo, R.E. (2017) *Masters of Craft: Old Jobs in the New Urban Economy*. Princeton, NJ: Princeton University Press.

Sennett, R. (2008) *The Craftsman*. New Haven, CT: Yale University Press.

Wilson, D. and Keil, R. (2008) The real creative class. *Social and Cultural Geography*, 9(8), 841–847.

8 Transient productions; enduring encounters

The crafting of bodies and friendships in the hair salon

Helen Holmes

"Are you going on holiday this year?" the polite, distant conversation filler of an awkward, yet intimate service encounter. A service encounter which may be somewhat prolonged: where you feel obliged to talk, to break the uncomfortable silence with the person working upon your body. It might be with the dentist, the doctor, a beautician, masseuse, or perhaps a hairdresser. In fact it is this very phrase which is often used to epitomise, even mock, the hairdresser–customer relationship. A relationship often viewed as inconsequential, fleeting and even fickle because of its everyday mundanity and potentially vain intentions.

In this chapter I explore the hairdresser as craft worker, moulding and making heads of hair, but crucially at the same time crafting friendships with clients. Hairdressing is very much classified as a typical service sector role: highly feminised, requiring minimal skill, and having little significance in people's lives. This portrayal of service work has been extensively critiqued by scholars over the past two decades, with calls for recognition and appropriate remuneration of the creativity, autonomy and often complex skillset required to conduct such service roles (see Bolton and Boyd, 2003; Gatta, 2009; Gatta et al., 2009). Yet service work is not aligned with craft or making. In response, I have contended elsewhere (Holmes, 2015a) that one reason service work is undervalued is because of the transient and intangible nature of the products it produces; the work of the hairdresser and the transience of the production of hair being a case in point. Rather, I argue that if we move beyond a focus on the outputs of service work, to a focus on the intangible, often invisible organisational, technical and cognitive skills (Hampson and Junor, 2005) involved in service roles, what emerges is a highly skilled, intuitive craft professional. A contemporary craft worker; moulding and shaping materials through repetitive craft practice. I now wish to extend this argument to contend that it is not just the labour which is undervalued in such craft service encounters but also the relationships such crafting creates. As I illustrate, through repetitive crafting and custom, friendships both blossom and endure.

Whilst friendship within craft and making has been a subject of academic interest, friendship within service work is a relatively under-researched subject. For instance, work examining dress-making (Hall and Jayne, 2016), stitch'n'bitch groups (Minahan and Wolfram Cox, 2007) and more broader forms of 'fabriculture' (Braitch and Brush, 2011) draw on the gendered and intimate friendship spaces created by these more normative acts of making. Yet friendship within forms of service work tends to focus on the relationships and camaraderie between employees (Casey, 2015; Hart, 2005). Only a limited number of studies address the relationship and bonds that can be formed between employees and customers (Cohen, 2010; Eayrs, 1993; Everts, 2010; Cranford and Miller, 2013). This limited research on friendship within service work is in contrast to the growing popularity of friendship more broadly as a topic of scholarly interest. The need to map intimacies and relationalities; and to recognise the varied spaces and contexts of friendship (Bunnell et al., 2012; Hall, 2009; Pahl, 2000; Sanger and Taylor, 2012) has become an increasing call. Consequently, this has led to a surge of studies on friendship ranging from debates around the flexibility of friendship (Spencer and Pahl, 2006; Rebughini, 2011); friendship as a supportive space for the marginalised (Askins, 2015; Hubbard, 2001; Weston, 1991); and as a means of identity formation and reciprocal value (Cronin, 2015). Other work has focused specifically on how materiality can structure and aid friendship (see Hall and Jayne, 2016; Karikari, 2014), whilst a few have considered how temporality functions within friend relationships (Cronin, 2015). Recognition of friendship as something which can be crafted through service work, and, in particular, through forms of service labour which involve crafting and making, has not occurred.

In what follows, I contend that friendship can be a key feature of employee–customer relations within service professions, and furthermore that such friendship is structured in part by the materiality and temporality of the products crafted. Drawing upon a year-long ethnographic study of a hair salon I argue that the only transient element of many hairdresser–customer service encounters is the production – hair. I explore how through the crafting of hair, friendships are also made, bound together in the transient palimpsest of hair. As I illustrate, such relationships can span decades, with their performance extending far beyond the space of the hair salon. Thus, hair becomes the temporal and material repository of not only the hairdresser's craft labour, but also of the relationship they craft with their customer.

The fleeting service encounter

Studies of service work, have often commented upon the fleeting nature of the service encounter, far removed from any form of activity we might refer to as craft. Fred Davis' 1950s study on cab driving deemed the relationship between the cab driver and 'his' fare as 'random, fleeting, unrenewable and devoid of socially integrative features' (1959: p.158). Jumping forward 30 years, this argument appeared to endure with service work labelled as responsible for the 'McDonaldisation' of society due to its scripted, routine and depersonalised nature (Ritzer, 1996). Added to this were debates around the anonymity and homogeny of post-modern spaces of consumption, in particular those which involved forms of service work. Marc Auge's (1995) notion of the 'non-place' – anonymous accelerated spaces where interactions remained faceless, and frequent – included all service work spaces: airports, supermarkets, shopping malls. Positioned against a backdrop of the intense and persistent debates from the 1970s regarding the deskilling of employment overall (see Braverman, 1974), it is little wonder that service work towards the turn of the century was perpetually positioned as unskilled, insignificant and thus requiring little pay or recognition. Indeed, the very antithesis to the highly creative, and skilled activities referred to as crafting.

Roll forward a decade and the 2000s saw debates surrounding service work changing considerably. Arlie's Hochschild's seminal piece on emotional labour (1984), the concept that workers had to manage their emotions to create organisational profit, had opened up the notion that service work did require skill (see also Hall, 1993; Lee-Treweek, 1997; Callaghan and Thompson, 2002). Furthermore, it exposed that such service work, and the 'soft' emotional skills required, was predominantly undertaken by women; hence its persistent position within the low paid, low respected ranks of employment. Emotional labour paved the way for recognition of a host of other interrelated forms of service labour, also requiring essentialised feminine attributes. Aesthetic labour (Witz, et al., 2003), whereby one embodies an organisation's ethos and branding through their appearance, was found to be at work within shop work (Pettinger, 2006), modelling (Entwistle, 2002) and flight attendants (Tyler and Abbott, 1998). Body work, whereby one performs paid work upon the bodies of others, such as care work (Twigg et al., 2011), beauty therapy (Sharma and Black 2001, Black 2002), hairdressing (Holmes, 2015a; Cohen 2010), alternative therapies (Oerton, 2004) and sex work (Sanders, 2005). This triplet of previously invisible forms of labour began to challenge dominant assumptions around the fleeting, unskilled and routine nature of service work. Yet, whilst worker creativity and physical and emotional forms of labour were finally being recognised, emphasis upon practical skill was still lacking. As a result, a number of studies began to critique the focus on emotional forms of labour within service work, arguing for a more in depth appreciation of the multifaceted 'blend of emotional, cognitive, technical and time management skills' such work involves (Hampson and Junor, 2005: p. 176; see also Gatta, 2009; Pettinger, 2006). Skills which are often so invisible, 'taken-for-granted' and mundane that they are overlooked (Junor et al., 2009).

In a previous piece of work (Holmes, 2015a), I built upon these debates to argue that the hidden and tacit skills of service work are a form of craft labour. Combining Richard Sennett's (2008) work on the physical and embodied act of craft practice, with Hampson and Junor's (2005) appeal to use fine-grained description to reveal the invisible skills of service work, I show that craft skills are present in professions, such as service roles, which do not produce stable, obdurate objects. In other words, craft is present in the practice as well as the end product. I illustrate how often in service work, such as hairdressing, beauty work, cleaning, and gardening, the product is temporary, thus the efforts of the service worker do not remain intact, rather they must be repeated on the same object over and over again. This repetitive craft practice operates on two levels. Borrowing again from Richard Sennett (2008), on the one hand, this is about the 'rhythm of routine' (p.268) and 'the ten thousand hours of experience' (p.20) repeating the same skill over and over until it becomes a form of proximal embodied knowledge (Hetherington, 2003). However, on the other, it is also about working on the same thing repeatedly – be it a specific head of hair, a particular lawn, or a certain house; it is something you get to know and understand the intricacies of. It is the latter which this chapter is interested in. In what follows, I use the example of the hairdresser, to illustrate that whilst service work craft labour reveals otherwise invisible skills due to the transient nature of the objects crafted, it is this transience which also reveals the relationships made between service worker and customer.

The palimpsest of hair

The hairdressing example used within this article, stems from 12 months conducting participant observation research while working in a hair salon in the north west of England. The salon which I refer to using the acronym Kirby's, was frequented predominantly by working-class women and was not considered 'high-end' (Yeadon-Lee et al., 2011) compared to other city or branded salons. During my time at Kirby's, I was responsible for sweeping up hair, making coffee, washing hair, answering the phone and cleaning. Being part of the everyday practices of the hair salon enabled me to appreciate and take stock of its networks, choreographies and relationships, whilst focus groups and interviews with salon staff, clients and friends of clients, enabled me to explore the ideas which were emerging. Right from the very beginning of my employment at Kirby's it was clear that hair was the linchpin holding everything together. Whilst this might seem obvious, given that it was a hair salon, the prominence of hair as a material substance, symbolic marker, or cultural reference point pervaded every practice, interaction and conversation within the space of the salon. Although biologically dead, and described as operating at the 'dead margins of the body' (Kwint 1999: p.9), hair was very much alive and potently present within the hair salon.

Materially hair was often described by both customers and stylists as having agency, doing as it pleased: going frizzy, getting greasy, sticking-up. This personification extended further with every head of hair described as unique and individual. As one participant, a former hairdresser, explained 'every head is different'. Such difference not only incorporates the biological material qualities of the hair – colour, texture, strength, curl – but also the layer upon layer of previous hair practices – colours, perms, even cuts. All of which are ascribed upon one head of hair, with some more prominent than others, as cuts grow out and the effects of perms and colours wear off. It is this inimitable, layered quality which has led me to refer to hair using the archaeological term 'palimpsest' (see also Holmes, 2014, 2015b). Originally used to describe 'a parchment or other surface on which writing has been applied over earlier writing which has been erased' (*Oxford English Dictionary*, 2005), contemporary definitions refer to the palimpsest as 'a superimposition of successive activities, the material traces of which are partially destroyed or reworked because of the process of superimposition' (Bailey, 2007: p.203). This contemporary definition accurately describes a head of hair; a personalised material record composed of unique biological features mingled together with the dominant to fading remains of previous hair practices.

Through the notion of the palimpsest of hair, the hairdresser's labour is revealed. No longer the transient fleeting crafting of an ever-changing object, but rather a labour stored in the individual head of hair, gradually ebbing away as time passes. This idea of craft labour being captured in the product it tends to is similarly reflected in the work of Caroline Steedman (2007: p.42), writing about the 18th century worsted spinner.

> She did not own the worsted on which she worked; it was never hers, neither in bundle form nor after she had exercised her energies on the bundles, mingled her labour with its greasy mass.... It was lent her, loaned her, given to her on trust, for payment that came only after the labour was complete.

Steedman's description of the worsted could easily be replaced with a head of hair, the spinner substituted with the hairdresser. As Susan Stewart (1999: p.30) discusses, 'materials store our labour and our maintenance'. Like the worsted, a head of hair becomes the material testimony of craft labours past. The notion of a crafted object signifying the work of a particular crafts person, either through hallmarking or recognisable style and design, is commonplace within studies of craft, particularly artisan crafts (see Goodsell, 1992; Sennett, 2008; Terrio, 1996). Yet, the key difference (aside from this being work upon a living body, not an inanimate object), as with most craft, is that unlike a brick or a Barolo goblet, the craft labour of the hairdresser is always fading.

Furthermore, the craft labour of the hairdresser is always competing with the remnants of other previous hairdressing labours. This may be their own work or that of others – including hairdressers and the customer themselves. Cutting is one such example, whereby previous cuts and styles impact upon future cutting practices, as Kirby's stylist, Alana, illustrates.

> Alana: This woman who I did the other week.... Anyway she had a bob and I could not follow it. I thought 'What has happened here?' So I said 'Where did you have your hair done before?' She said 'I always have my hair done here.' .... She said Sophie had done it, 'Then Beth did it. But I

loved it when Clare did it last time, it was absolutely gorgeous!' And I thought I'm going to have to re-cut all this because I can't follow it.

(Salon Focus Group)

Alana's quote illuminates how the remnants of past hair practices can cause problems for future ones. In this case, the customer's hair has a pattern which Alana cannot follow, thus the craft labour of other hairdressers is stored in the customer's hair, affecting what happens to it next. Figure 8.1 shows a customer getting their hair cut.

Colouring is another example where previous labours become stored in the palimpsest of hair. The following fieldwork diary excerpt explores the havoc previous hair colours can wreak on future ones.

If a customer wants to go, say blonde, but the colour on their hair is darker, then they have to have it stripped with bleach. When this is done bands of colour appear because at the top of the head, near the root, the hair is more porous because it has had less colour applied to it in the past (as it is newer hair), therefore, the bleach/peroxide combination works easily there. However, as you move down the hair, where more layers of colour have been applied in the past (because it's older hair), then the

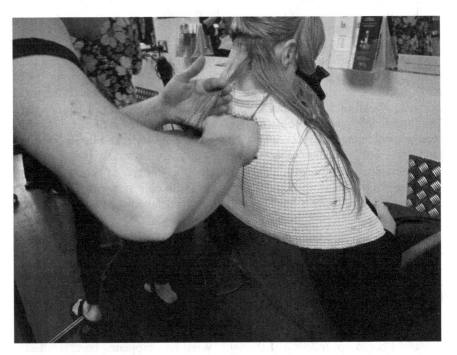

Figure 8.1 Cutting hair
Source: Photograph by Helen Holmes

stripping doesn't work as well and, thus, sends it either red or ginger; the nearer the ends of the hair (the oldest hair), the least the effect.

(Fieldwork Diary Extract)

At Kirby's there were numerous cases of customers visiting the salon, who had attempted to colour their own hair and something similar to the above, or worse, had happened; bands of graduated colours appearing from root to tip because of the lingering effects of other colour practices. One such case involved a woman who had tried to dye her dark brown hair blonde at home and it had turned orange, and she had visited the salon in tears to try to get it sorted. Although Kirby's stylist, Linney, was more than happy to oblige, later on, away from the salon floor, she commented to myself and another stylist: 'Well that's what you get when you mess with your own hair.' Whilst these examples corroborate the idea of previous craft labours being stored in the palimpsest of hair, they also raise the question of who has the right to craft the hair?

For Steedman's spinner the worsted was only ever 'loaned'. However, the ownership of and right to labour on hair appears somewhat blurred. Linney's comment suggests that certain hair practices, such as colouring, should only ever be done by a professional and that customers should leave well alone of their own hair. Thus, there is a form of stewardship (see Lane and Watson, 2012) occurring over the product of hair, whereby the hair is owned by one person (the wearer) but the management and making of it is conducted by someone else (the hairdresser). However, this is complicated further when one considers participants' guilt at being seen by their hairdresser, outside the space of the salon, with less than 'perfect' hair.

> Cos I get my eyebrows done in the same place, and there's a few times I've gone in and just felt ashamed because my hair's been that much of a mess. And I always feel like I have to apologise to her.... cos they're trying to make it look nice, and you think I'm not doing very well here, or a good job of keeping up what you've done.
>
> (Jennifer, Interview)

> Erm you feel a bit guilty if they've spent ages doing it nice and then you have it scrapped back.
>
> (Eileen, Interview)

These quotes illuminate a partnership of shared ownership/stewardship of the product of hair. The participants recognise that their hair stores the craft labour of the hairdresser, and they try to maintain it. Yet, there is friction around this maintenance and repair, and the stewardship/ownership boundaries are not clearly defined. As Stevenson (2001: p.149) concludes, the hairdresser 'creates work for the consumer in their daily production and maintenance of the style, as well as future work for themselves'. Thus, hair is co-produced, made between hairdresser and client; with the labour of the customer mingled in with the craft labour of the hairdresser; both in some ways reliant on the other, but neither one having complete control. Russell Belk (2010: p.718) writes of commodity exchange as being 'ideally... simultaneous', goods in exchange for money, 'so that there is no lingering debt to tie the parties together'. Yet, as the above has shown, this is not the case with hair, as something of the hairdresser's craft is always retained in the product. Belk continues that the commodity exchange should be about 'the reproduction of rights to objects, not the reproduction of relationships between people'. However, the transience of hair demands its repeated crafting, thus the relationship between hairdresser and customer is consistently reproduced, as is the object of exchange. It is to this repetitive reproduction that we now turn, firstly exploring the intimacy created by these repeat craft encounters.

The intimate encounter

The previous section illustrated how the transient character of the palimpsest of hair binds the customer to the hairdresser for repeat crafting. Of course, there is the argument that some people never visit the same hairdresser, opting to move around, or maybe never visit one at all. However, many people choose to stay for a prolonged period of time with one stylist. Before discussing the types of relationships this repetitive exchange can create, particularly friendships, I wish to focus on the embodied nature of such encounters, and how this embodiment also binds hairdresser to customer. As I have already illustrated, the material of hair is crafted in the hair salon, thus bodies are crafted in the salon. This notion of body work taking place within the salon is not new (see Cohen, 2010; also Sharma and Black, 2001; Black, 2002 for a discussion on beauty salons; and body work more generally Wolkowitz, 2002). Body work typically is enthused with notions of caring and compassion, where a worker, nearly always female, attends to the bodily needs of a customer. Sharma and Black (2001) make the connection that the body work performed in the salon is as much about working on identities as it is bodies, and the two cannot be seen as distinct (see also Stevenson, 2001). Similarly, Toerien and Kitzinger (2007) explore how beauty therapists must navigate between attending to a client's emotional needs, whilst also attending to their body. However, whilst these salon-based accounts are undoubtedly grounded in the corporeal elements of work, the centrality of labour somewhat overshadows the embodied intimacy of such encounters.

In her work on domestic friendships and motherhood, Anne Marie Cronin (2015) uses the term 'inclusive intimacy' to describe friendship groups which unite around the 'common connection' (p.669) of having children. Adapting this term, I argue that the relationship between hairdresser and client can be one of 'inclusive embodied intimacy', made possible by the transient palimpsest of hair requiring repeated crafting and the hairdresser's exclusive knowledge of it and their client. This is best illustrated by two quotes from participant's Samantha and Jennifer.

> Yeah cos they do get to know your hair and they get to know you and it's a bit of a unique relationship isn't it! A very trusting one!
>
> (Jennifer, Interview)

> Because every time anybody else has cut my hair, I have hated it. But Heather [another hairdresser] has explained to me that it's because they need to factor cutting **your** hair... and Diana [Samantha's hairdresser] is used to **my** hair.
>
> (Samantha, Focus Group)

These quotes raise several points. Firstly, and as both participants stress, to be able to properly craft someone's hair requires knowledge of that hair and its distinctive material properties, in other words the palimpsest. Secondly, the embodied nature of the interaction is dyadic: the client's hair/body is crafted by the hairdresser, but only through the skills inherent in the hairdresser's body. The hairdresser stores their knowledge and labour in the client's hair, whilst simultaneously learning about the client's hair through the craft process. Thus, the tacit knowledge of one body becomes embodied in that of the other through an iterative craft practice. This is captured in Samantha's quote about visiting another hairdresser and him/her not having the embodied skill or knowledge to deal with her hair, as she likes. As Jennifer stresses, having the right knowledge about another's hair and their preferences with it, yields a 'unique relationship'; made inclusive because of the inimitable material of hair binding the two bodies together. Thirdly, and crucial to both of the above points, is trust.

Trust was mentioned by numerous participants as an essential attribute of the hairdresser/client relationship. As one participant, Katherine, described during an interview, 'hairdressing is about trust because there's nothing worse than a girl with a bad haircut'. This focus on trust highlights the anxieties around surrendering one's hair (and body) over to someone else. Such apprehensions were often played out in the salon when someone else, such as a salon junior, took over managing a client's hair.

> I hate it. I end up saying 'Just condition it'. I just don't like people touching it... obviously having your hair cut they do need to touch it. But it's always like the junior member of staff washing it.
>
> (Carla, Focus Group)

> I'm constantly sitting there going 'Ouch! Ouch', When they feel the need to brush your hair afterwards as well, the hair wash people.
>
> (Lena, Interview)

As the quotes convey, the salon junior performing the menial task of hair washing is the unknown entity. Devoid of the hairdresser's embodied knowledge of the palimpsest through repeat craft practice, her[1] touch feels alien, inadequate to the client's corporeal needs. Her involvement bursts the embodied intimate customer–hairdresser bubble, creating discord for the client. She has not built up the trust that the hairdresser has, nor has she had the opportunity to gather embodied knowledge of the customer's hair. Figure 8.2 illustrates the embodied nature of having your hair cut at the salon. However, 'trusting' your hairdresser extends much further than just trusting her with your hair. As the following section illustrates, just as bodies and identities are crafted in the hair salon, so are friendships.

The crafting of friendship

I have already discussed how the transient nature of the palimpsest of hair pushes the client back to the salon for repeated crafting, crafting which creates a particular type of inclusive embodied intimacy due to the imitability of the material of hair. In addition to the repetition of the encounter, it must be noted, that each appointment may take several hours depending on what is being done. For example, colouring or perming appointments generally take

Figure 8.2 The embodied hair washing encounter
Photograph by Helen Holmes

longer than just cutting or blow waving ones, due to the intricacy of the practices and the involvement of chemicals which need time to 'take'. This means that not only are visits to the hairdresser repetitive but that each craft encounter is a lengthy one-to-one interaction. There are several studies examining the interactive characteristics of the hairdresser–client appointment, the majority of which argue that hairdressers, as part of emotional labour, are expected to engage their client in conversation (Cohen, 2010; Gimlin, 1996; Kapp Howe, 1977). In other words, this interaction takes place to ensure the commodity exchange. Louise Kapp Howe (1977) discusses how hairdressers ignore what their clients are saying, whereas Deborah Gimlin (1996) argues that they only listen to keep the customer happy and paying. Rachel Lara Cohen (2010) takes this one stage further, contending that the extent to which the hairdresser is willing to engage with their clients depends on their role. So salon owners who 'have the most to lose' (p.202) will go out of their way to do 'favours' (p.207) and seemingly befriend clients; whilst 'waged stylists have little incentive to befriend clients' (p.214) because it makes little difference to their job or earning capacity. These arguments I disagree with.

I now return to the notion of trust. So far I have discussed how one must trust the hairdresser to craft their body as they desire. However, trust also extends to the interactive and communicative elements of hairdressing. The notion of the hairdresser or beauty worker as a therapist whom clients confide in and share intimate details about their lives with is not new (see Gimlin, 1996; Sharma and Black, 2001). Indeed, Parr, a stylist at Kirby's, would regularly wear a t-shirt to work emblazoned with 'Therapist' across the front. In a noteworthy piece on black hair salons, Adia Harvey-Wingfield (2009) describes how the salon is a 'safe space for black women' (p.80) where 'no topic of conversation is off limits' and 'their concerns, issues and perspectives were fundamental rather than marginalized or ignored' (p.83). But playing the role of therapist, or creating a safe space where people feel relaxed to talk about their lives and their views is not simply about performing emotional labour to complete a commodity exchange. Creating such relationships requires trust and that trust takes time to cultivate: time which is given by the lengthy, repetitive, and inclusive embodied intimacy of the crafting encounter. Such trust may only extend as far as making the hairdresser an acquaintance or a 'limited friend' (Spencer and Pahl, 2006), but likewise it may also lead to something more, a genuine friendship.

Paola Rebughini (2011: 2.13) talks about how trust is an important element in maintaining inter-subjective relationships, but in friendships it performs a 'distinctive' role, granting stability to the relationship. Furthermore, she draws on the work of Simmel (1908) to argue that discretion and disclosure are also necessary elements of the intimate structure of friendship (2.19). I cannot dispute the idea that the salon is a safe space for clients to reveal their secrets and intimate thoughts. As salon owner Patricia noted when talking about clients' disclosures 'if I told everything that people had told me I'd be in a straight-jacket'. Yet such exchanges are not simply one way, as the majority of salon and service-based literature would lead us to believe. Over time friendships are crafted and these are not always biased towards one party, but can involve reciprocal and mutual forms of intimacy. Some of these friendships span decades, in the case of Kirby's there were clients who had been attending the salon for over 30 years. The following fieldwork diary recounts an exchange between client Dot and owner Patricia during an appointment.

> Today I sat in on a conversation between Patricia and Dot, whilst Patricia set her hair. The talk centred upon their families and what they were doing. Patricia mentioned how she was concerned about her son going off to university, whilst Dot spoke of her ill health. Therefore, although the relationship has an exchange value, in terms of Dot paying Patricia to do her hair, it is also based upon mutual friendship. They each share what they have been up to and the hairdressing experience offers them a chance, as friends, to catch up.
>
> (Fieldwork Diary)

Dot has been a client of Patricia's for nearly 30 years and as the conversation illuminates the exchange is not one way, but rather is structured by disclosure and discretion between the two as they catch up about each other's lives.

Another example, from participant Nancy, highlights how client–hairdresser friendships can be structured around the common connections made between the two (Spencer and Pahl, 2006; Cronin, 2015).

> Like I will go once every six weeks now and she remembers ABSO-LUTELY EVERYTHING that I've told her.... We have the same opinions about things... we have like really in depth conversations, like about politics and you know what I mean.... It's my 30th next year and I would definitely invite her. And it's not like a loyalty thing. I think it's just a simple fact that I really like her. I really, really like her and she's just such a lovely person.
>
> (Nancy, Interview)

As Nancy describes, her and her hairdresser Debbie bond over their similar opinions; making their conversations both in depth and memorable. Furthermore, Nancy's quote illustrates another important feature of hairdresser–client friendships – they do not solely exist within the space of the salon. Nancy's wish to invite Debbie to her 30th birthday party is not unusual. Invites to client's weddings, birthdays, funerals, and exchanging gifts are part of the course for the hairdresser (see Cranford and Miller, 2013 for a similar discussion on careworkers). Whilst I was working in the salon everyone was invited to the engagement party of Denise, one of the regular clients; and stylists commonly talked of meeting a few regular clients for a drink outside of work. Thus, hairdresser–client relationships demonstrate many features of friendship, and I argue in some circumstances that is just what they are.

Indeed, on the flip side of these intimate hairdresser–client friendships are people who are envious of such relationships. Participant, Eileen, who regularly changed hairdresser, exclaimed that she was 'quite envious of people who have this rapport going with their hairdresser'. Whilst another participant, Vanessa, expressed anxiety at the 'massive conversations' other people seem to have with their hairdressers. She questioned whether she should 'be jealous of this hairdresser/person relationship' and 'having a lot more involvement'. Vanessa and Eileen's envy, highlights the social relevance of these hairdresser–client friendships, despite their overlooked and often disputed position within service work literature. As stylist Parr summarises:

> I do think that you build up a relationship with your client... They can go anywhere and just have a trim, but they come back to you because you build-up a friendship with them.
>
> (Parr, Salon Focus Group)

As Parr's quote confirms, hairdressers create relationships with their clients over time because of the repetitive nature of the craft encounter. They could go anywhere for a trim, as Parr notes, but they are unlikely to because of the intimacy they share with him. An intimacy which can be so inclusive and embodied that the boundaries of ownership of the client's hair become blurred; with both stylist and client confiding in each other about their innermost thoughts and sharing the highs and lows of life in and outside the space of the salon.

Conclusion

This chapter has emphasised the multiplicity of craft and making. Through the example of the hairdresser and the hair salon, it has revealed how hair is crafted within the salon; however, so too are relationships between hairdresser and client. Drawing on previous work which argued that the invisible skills of service work can be akin to a form of repetitive craft practice, I have argued that service roles can extend further than crafting a transient product, but in turn can also craft intimate relationships. The hairdressing example used has illuminated how through repetitive interactive encounters between service workers and customers, relationships and even friendships can be produced, bound together by the temporality and materiality of the products crafted. Yet, whilst the products of service work may be transient, unstable and fleeting, I contend that both the skills of the service workers and the relationships they craft with their clients are not.

As regards hairdressing, such relationships are bound indisputably with the palimpsest of hair. Hair's inimitable qualities ensure a co-productive relationship between client and hairdresser, where the boundaries of ownership become blurred, as both store their labour in the material. Through repeat and often lengthy appointments, this unity is extended through what I have termed inclusive embodied intimacy; whereby the bodies of hairdresser and client are connected through iterative craft practices upon the palimpsest of hair. The hairdresser storing their knowledge and craft labour in a client's hair, whilst simultaneously learning further embodied knowledge about that head of hair in the process. All of which is bound up with trust: a bodily trust, which often only becomes apparent when someone else touches a client's hair, but also a form of communicative and interactive trust. To all appearances, this trust may be presented as one way, superficial – the hairdresser listening to the intimate disclosures of the client, performing their role as emotional labourers. However, further probing may reveal something much more meaningful; two way reciprocal interactions between client and hairdresser. As I have illuminated, such interactions can and do display features of friendship from disclosure and discretion to finding common connections. These relationships extend outside the space of the salon and the formalities of the commodity exchange. Thus, relationships do exist between hairdresser and client which are more than just a commodity exchange or a performance of emotional labour; rather they can be enduring, genuine and significant friendships.

I do not refute that there will be many people who do not have a close relationship with their hairdresser. Likewise, there will be some salons where relationships may remain much more formal (see Yeadon-Lee et al., 2011 and Chugh and Hancock, 2009 for discussions on high-end salons). More broadly, there will be types of service roles, particularly those without repetitive or lengthy encounters with customers, which are less likely to result in friendship. Rebughini (2011) talks of friendship as 'an elastic and negotiable interpersonal space'; relationships and the interactions they create are not fixed. Thus, some service worker–customer relationships may only be on the level of acquaintance, some may be 'limited friendships' (Spencer and Pahl, 2006), or 'paid friends' (Eayrs, 1993: p. 20), others may be much more. However, as this chapter has demonstrated, service work relationships are far more complex and multifaceted than simply being about exchanging a commodity or performing a particular form of labour. They are not fleeting encounters, involving unskilled workers and faceless customers and employees. Rather service work can and does involve craft and making, and that craft extends not just to the often invisible and intangible skills involved, or to the transient and temporary products and services created, but also to the intimate social relations such encounters can make.

Note

1 I refer to the salon junior as 'her' as hairdressing is a predominantly female occupation. As recent industry statistics illustrate 90% of the work force is female (Habia, 2011/12).

References

Askins, K. (2015) Being together: everyday geographies and the quiet politics of belonging. *ACME: An International E-Journal for Critical Geographies*, 14(2), 461–469.

Auge, M. (1995) *Non-places: Introduction to an Anthropology of Supermodernity.* London: Verso.

Bailey, G. (2007) Time perspective, palimpsests and the archaeology of time. *Journal of Anthropological Archaeology*, 26(2), 198–223.

Belk, R. (2010) Sharing. *Journal of Consumer Research*, 36(5), 715–734.

Black, P. (2002) Ordinary people come through here: locating the beauty salon in women's lives. *Feminist Review*, 71, 2–17.

Bolton, S.C. and Boyd, C. (2003) Trolley dolly or skilled emotion manager? Moving on from Hochschild's *Managed Heart. Work, Employment & Society*, 17(2), 289–308.

Braitch, J.Z. and Brush, H.M. (2011) Fabricating activism. *Utopian Studies*, 22(2), 233–260.

Braverman, H. (1974) *Labor and Monopoly Capital: The Degradation of Work in the Twentieth Century.* New York: Monthly Review Press.

Bunnell, T., Yea, S., Peake, L., Skelton, T. and Smith, M. (2012) Geographies of friendship. *Progress in Human Geography*, 36(4), 490–507.

Callaghan, G. and Thompson, P. (2002) 'We recruit attitude': the selection and shaping of call centre labour. *Journal of Management Studies*, 39(2), 233–254.

Casey, E. (2015) 'Catalogue communities': work and consumption in the catalogue industry. *Journal of Consumer Culture*, 15(3), 391–406.

Chugh, S. and Hancock, P. (2009) Networks of aestheticization: the architecture, artefacts and embodiment of hairdressing. *Work, Employment and Society*, 23(3), 460–476.

Cohen, R.L. (2010) When it pays to be friendly: employment relationships and emotional labour in hairstyling. *The Sociological Review*, 58(2), 197–218.

Cranford, C. and Miller, D. (2013) Emotion management from the client's perspective: the case of personal home care. *Work, Employment and Society*, 27(5), 785–801.

Cronin, A.M. (2015) Domestic friends: women's friendships, motherhood and inclusive intimacy. *The Sociological Review*, 63, 662–679.

Davis, F. (1959) The cabdriver and his fare: facets of a fleeting relationship. *The American Journal of Sociology*, 65(2), 158–165.

Eayrs, M.A. (1993) Time, trust and hazard: hairdressers' symbolic roles. *Symbolic Interaction*, 16(1), 19–37.

Entwistle, J. (2002) The aesthetic economy: the production of value in the field of fashion modelling. *Journal of Consumption Culture*, 2(3), 317–339.

Everts, J. (2010) Consuming and living the corner shop: belonging, remembering, socialising. *Social & Cultural Geography*, 11(8), 847–863.

Gatta, M., Boushey, H. and Appelbaum, E. (2009) High touch and here-to-stay: future skills demands in US low wage service occupations. *Sociology*, 43(5), 968–989.

Gatta, M. (2009) Balancing trays and smiles: what restaurant servers teach us about hard work in the service economy. In Bolton, S. and Houlihan, M. (eds.) *Work Matters: Critical Reflections on Contemporary Work*. Basingstoke: Palgrave-Macmillan.

Gimlin, D. (1996) Pamela's place: power and negotiation in the hair salon. *Gender and Society*, 10(5), 505–526.

Goodsell, C.T. (1992) The public administrator as artisan. *Public Administration Review*, 52(3), 246–253.

Habia (2011/12) *Industry Overview*. Available from: http://www.habia.org/industry/overview [Accessed 15th June 2016].

Hall, E. (1993) Smiling, deferring & flirting: doing gender by giving 'good service'. *Work and Occupations*, 20(4), 452–471.

Hall, S.M. (2009) 'Private life' and 'work life': difficulties and dilemmas when making and maintaining friendships with ethnographic participants. *Area*, 41(3), 263–272.

Hall, S.M. and Jayne, M. (2016) Make, mend and befriend: geographies of austerity, crafting and friendship in contemporary cultures of dressmaking. *Gender, Place and Culture*, 23(2), 216.

Hampson, I. and Junor, A. (2005) Invisible work, invisible skills: interactive customer service as articulation work. *New Technology, Work and Employment*, 20(2), 166–181.

Hart, E. (2005) Anthropology, class and the 'big heads': an ethnography of distinctions between 'rough' and 'posh' amongst women workers in the UK pottery industry. *The Sociological Review*, 53(4), 710–728.

Harvey-Wingfield, A. (2009) *Doing Business with Beauty: Black Women, Hair Salons and the Racial Enclave Economy*. Plymouth: Rowan & Littlefield.

Hetherington, K. (2003) Spatial textures: place, touch and praesentia. *Environment and Planning A*, 3(11), 1933–1944.

Hochschild, A.R. (1983) *The Managed Heart: Commercialization of Human Feeling*. London: University of California Press.

Holmes, H. (2014) Chameleon hair: how hair's materiality affects its fashionability. *Critical Studies in Fashion & Beauty*, 5(1), 95–110.

Holmes, H. (2015a) Transient craft: reclaiming the contemporary craft worker. *Work, Employment & Society*, 29(3), 479–495.

Holmes, H. (2015b) Self time: the importance of temporal experience within practice. *Time & Society*. doi:10.1177/0961463X15596461

Hubbard, P. (2001) Sex zones: intimacy, citizenship and public space. *Sexualities*, 4, 51–71.

Junor, A., Hampson, I. and Ogle, K.R. (2009) Vocabularies of skill: the case of care and support workers. In Bolton, S. and Houlihan, M. (eds.) *Work Matters: Critical Reflections on Contemporary Work*. Basingstoke: Palgrave-Macmillan:

Kapp Howe, L. (1977) *Pink Collar Workers: Inside The World of Women's Work*. New York: Avon Books.

Karikari, I. (2014) Casserole Club. *International Journal of Integrated Care*, 14(8), Conference Abstract.

Kwint, M. (1999) Introduction. In Kwint, M., Breward, C. and Aynsley, J. (eds.), *Material Memories*. Oxford: Berg.

Lane, R. and Watson, M. (2012) The stewardship of things: property and responsibility in the management of manufactured goods. *Geoforum*, 43(6), 1254–1265.

Lee-Treweek, G. (1997) Women, resistance and care: an ethnographic study of nursing auxiliary work. *Work, Employment & Society*, 11(1), 47–63.

Minahan, S. and Wolfram-Cox, J. (2007) Stitch'n'bitch: cyberfeminism, a third place and the new materiality. *Journal of Material Culture*, 12(1), 5–21.

Oerton, S. (2004) Bodywork boundaries: power, politics and professionalism in therapeutic massage. *Gender, Work & Organisation*, 11(5), 544–563.

Oxford English Dictionary (2005). *Oxford English Dictionary*. Oxford: Oxford University Press.

Pahl, R. (2000) *On Friendship*. Cambridge: Polity Press

Pettinger, L. (2006) On the materiality of service work. *The Sociological Review*, 54(1), 48–67.

Rawlins, W.K. (2009) *The Compass of Friendship: Narratives, Identities, and Dialogues*, Thousand Oaks, CA: Sage.

Rebughini, P. (2011) Friendship dynamics between emotions and trials. *Sociological Research Online*, 16(1), 3. Available from: http://www.socresonline.org.uk/16/1/3. html [Accessed 15th June 2016].

Ritzer, G. (1996) *The McDonaldisation of Society*. London: Sage.

Sanders, T. (2005) *Sex Work: A Risky Business*. Devon: Willan Publishing.

Sanger, T. and Taylor, Y. (2012) *Mapping Intimacies: Relations, Exchanges, Effects*. London: Palgrave Macmillan.

Sennett, R. (2008) *The Craftsman*. London: Penguin.

Sharma, U. and Black, P. (2001) Look good, feel better: beauty therapy as emotional labour. *Sociology*, 35(4), 913–931.

Simmel, G. (1908) *Soziologie*, Leipzig: Duncker & Humbolt. Cited in Rebughini, P. (2011) Friendship dynamics between emotions and trials. *Sociological Research Online*, 16(1), 3. Available from: http://www.socresonline.org.uk/16/1/3.html [Accessed: 20th May 2016].

Spencer, L. and Pahl, R. (2006) *Rethinking Friendships: Hidden Solidarities Today*. New Jersey: Princeton University Press.

Steedman, C. (2007) *Master and Servant: Love and Labour in the English Industrial Age*. Cambridge: Cambridge University Press.

Stevenson, K. (2001) Hairy business: organising the gendered self. In Halliday, R. and Hassard, J. (eds.) *Contested Bodies*. London: Routledge.

Stewart, S. (1999) Prologue: from the museum of touch. In Kwint, M., Breward, C. and Aynsley, J. (eds.) *Material Memories*. Oxford: Berg.

Terrio, S. (1996) Crafting grand cru chocolate in contemporary France. *American Anthropologist*, 98(1), 67–79.

Toerian, M. and Kitzinger, C. (2007) Emotional labour in action: navigating multiple movements in the beauty salon. *Sociology*, 41(4), 645–662.

Twigg, J., Wolkowitz, C., Cohen, R.L. and Nettleton, S. (2011) Conceptualising body work in health and social care. *Sociology of Health & Illness*, 33(2), 171–188.

Tyler, M. and Abbott, P. (1998) Chocs away: weight watching in the contemporary airline industry. *Sociology*, 32(3), 430–450.

Weston, K. (1991) *Families We Choose: Lesbians, Gays and Kinship*. New York: Columbia University Press.

Witz, A., Warhurst, C. and Nickson, D. (2003) The labour of aesthetics and the aesthetics of organisation. *Organization*, 10(1), 33–54.

Wolkowitz, C. (2002) The social relations of body work. *Work, Employment, Society*, 16(3), 497–510.

Yeadon-Lee, T., Jewson, N., Felstead, A., Fuller, A. and Unwin, L. (2011) Bringing in the customers: regulation, discretion and customer service narratives in upmarket hair salons. *International Journal of Interdisciplinary Social Sciences*, 6(3), 101–114.

9 Entangled corporeality in the making of taxidermy

Elizabeth R. Straughan

Introduction

> At the beginning it was really hard, and, you try to explain to somebody how hard it actually is to stuff that thing and, I am sure you know [pause] it is really hard to understand what is going on when you are taking the skin off. I think that is the biggest change, like learning, making my body understand it... before I would just look at that and just see skin with a lump and not understand, but now I know exactly when I have to push and where. And it's just, through doing it repetitively.
>
> (Jasmine, pers. com. 15/06/2010)

In this interview transcript, artist cum taxidermist Jasmine alludes to the challenges of learning to undertake the practice of taxidermy. She highlights the importance of her own fleshy body in the process of making a taxidermy mount and the need to make it understand the materiality with which she engages. This chapter unpacks this process through consideration of the haptic qualities pertaining to this practice, qualities that facilitate the generation of tacit knowledge (Polanyi, 2009[1966]) about skin, flesh and bone. That is, I attend to tactile and muscular acts that require development in the process of *learning* taxidermy.

The word taxidermy is derived from the Greek words *taxis* meaning order or arrangement, and *derma*, translated as skin (Harper, 2017). In more literal terms taxidermy means the arrangement of skin. Creating a convincing piece of taxidermy involves the careful process of removing skin from an animal's body, preparing that skin so that it does not decay over time, and preparing a mount onto which that skin may be arranged. The aim is to produce a mount that mimics an animal in life, and to do so a taxidermist must devote time and effort in crafting a form that fully references the fleshy body of the individual animal upon which he or she is working (Brown, 2013).

Taxidermy is a practice characterised by proximity and intimacy (Straughan, 2015), whereby an in-depth understanding of the body of another is achieved via corporeal contact, or as Jasmine explains, via the process of "learn[ing] things with your hands" (pers. com. 15/06/2010). Such a process is predicated on the sense of touch – the ability of the hands to feel and work with texture, tensions and frictions – and the capacity of muscles and tendons in the human hand and arm to pull, push and twist. It requires the use of 'internal' or somatic sensations comprised of proprioception, kinesthesia, and the vestibular system; proprioception relates to felt muscular movement, kinesthesia to the sense of movement itself, while the vestibular system is responsible for balance (see, for example, Dixon and Straughan, 2010; Paterson, 2009). Touch is a sense that draws attention to the import of both the internal and external aspects of the body in sensing and making sense of the world. An imaginary of touch, then, is one of immersion and entanglement (Hawkins and Straughan, 2014).

Drawing on the 'hands on' material qualities of taxidermy practice this paper looks to the ways in which dead and alive bodies can become entangled in the process of making a mount. In so doing it extends discussions on the geographies of making by not only exploring the skills of making taxidermy mounts, but also the role of the material worked on. That is, in this chapter I look to the micro-encounters of taxidermy practice (Carr and Gibson, 2016) to consider the role of non-living, non-human materials and their capacity to affect the making process. Further, I offer a materially focused contribution to those post-human arguments that challenge isomorphic conceptions of the human subject, and appreciate the human as emergent *with* the world (for example see Wylie, 2005 and 2006). Looking to the work of Irigaray (1993) and Bennett (2010) I consider the process of making taxidermy through a feminist materialist lens to acknowledge the vibrancy of dead flesh in an assemblage, a vibrancy *with* which taxidermists work. Paying attention to the more-than-human aspects of taxidermy practice, I argue that this is a creative process through which corporeal entanglements emerge. To make this claim, I focus on the effects and role of dead vibrant matter on the maker and the creative process, before inverting the focus to consider the tensions and corporeality of the human maker's body on the dead animal's body.

To relay the corporeal entanglements at play in taxidermy practice this chapter draws on film transcripts taken from practice-based sessions that saw me doing and observing taxidermy, audio based interview transcripts, and audio based transcripts taken from my initial taxidermy lesson with an experienced taxidermist. However, before a more detailed discussion of this methodology, I consider the role touch can play in a vital materialist approach to exploring engagement between live human and dead non-human bodies in making practices.

From animal geographies to vital materiality

Geographers Hodgetts and Lorimer (2015) have recently argued that there has been a neglect of '*animal's geographies*' (emphasis in original) within the discipline, with regards to animal lives in the field. This is a neglect born in part, they suggest, from a lack of methodological tools that enable geo-graphers exploring non-human animals to do so in a manner that is not human centred. In this chapter, I take up these geographers' call to extend our methods in the pursuit of studying animals' geographies by presenting my own experimental, embodied and sensory focused project that drew out the role of both human *and* non-human bodies in the creation of taxidermy mounts. To do so, I focus on the sense of touch through a vital materialist approach that pulls out the capacity of *both* dead and living bodies to affect the creative process.

Attentiveness to agency and vitality in the study of non-human animals has, Lorimer and Driessen explain, emerged within a bio-political turn that has unpacked the co-mingling of heterogenous bodies through a "vital materialist concern for the lively potentials of nonhuman forms and processes" (2013: p.250). Embracing the variety of forms that living matter takes, geographers looking to developments in biotechnology that have raised the question of what counts as life have positioned non-corporeal living matter as a valid object of inquiry (Davies, 2003 and 2006; Dixon, 2008 and 2009). In this endeavour, matter is itself a complex subject of inquiry for, as Braidotti states, "a [b]iocentered perspective affects the very fibre and structure of social subjects" (2010: p.201). This perspective, located within a Spinozist framework, draws attention to the "mutual interdependence of material, biocultural and symbolic force in the making of social and cultural practices", and thus a methodological focus on process and interconnection (Braidotti, 2010: p.203).

Elsewhere, Bennett (2010) has made a strong case for an appreciation of 'dull matter' as vital when in an assemblage. Building a theory of *distributive agency*, Bennett takes her lead from Spinoza to argue, "bodies enhance their power in or as a heterogeneous assemblage" (2010: p.23). That is, the affectivity of agency is distributed across a field of both 'beings' *and* 'things,' rather than being a capacity localised in a human body or in a collective produced only by human efforts. Here, 'things' or objects are considered capable of creating effects when they combine with other 'vitalities', whether human or non-human. Within an assemblage, the effects generated are "emergent properties, emergent in that their ability to make something happen (a newly inflected materialism, a blackout, a hurricane, a war on terror) is distinct from the sum of the vital force of each materiality considered alone" (Bennett, 2010: p.24). Each vitality, both of matter and life, exerts its own agentive force, separate from but also shaping the overall effect that the assemblage has.

To pay attention to the vital materiality of dead matter in the making of taxidermy, I look to the sense of touch as it enables a methodological focus on both process and interconnection (Braidotti, 2010). Touch produces an imaginary that "speaks to our exposure to, and immersion in, the world of other…" (Shildrick, 2001: p.402). That is, this sense signals the porosity of the body, a concept taken up by Luce Irigaray's work on feminist materialism. As Fullagar elaborates, in Irigaray's lexicon, porosity dissolves any straightforward notions of inside and outside as it "situates the body as a threshold or passage, not strictly defined by inside/outside but rather by multiple surfaces open to other surfaces" (2001: p.297). Touch and the porosity it enables provides a means to consider the emergence of inter-connectivity between two bodies. In the next section, I discuss the methodological practice used to draw attention to the porous relations between dead and living bodies.

Methods to explore skin that is moved

Geographic enquiry into taxidermy is not new to the discipline. Patchett (2008) has considered a processual perspective of taxidermy mounts to argue that their entanglement in various human networks enables their 'after-life'. In collaboration, Patchett and Foster (2008) together have drawn out the invisible work of taxidermy making and maintenance to challenge the focus on representational surface, instead describing the "existing remnants of the animal" [i.e. the Blue Antelope] (2008: p.104) – skin, bone and archival material – in the form of drawings, scientific papers and other historical remains. Their re-construction considered the historical, political and cultural contingencies that went into the process of making. More recently, Patchett (2012) has written about her taxidermy apprenticeship, where her own body became an instrument for research to gain a sensate understanding of this practice, alongside the use of film (Patchett, 2015).

My own research into taxidermy formed part of a wider enquiry into the geographies of skin, wherein I was interested in the role of touch (the locus for which is the skin) and the 'somatic senses' (Paterson, 2007) in the experience of practicing taxidermy. On commencing this project, creativity was not planned within the initial, proposed methodology, which was mainly driven by methods dominated by talk and text. However, during the early stages of the research I was influenced by feminist arguments for the use of 'supplication', an approach informed by an understanding that the interviewer, "seek[s] reciprocal relationships based on empathy and mutual respect… often sharing their knowledge with those they research" (England, 1994: p.243). Alongside this, I was concerned with the question of how to craft and ask interview questions about the tactility and haptics of a practice with which I had never engaged. To overcome this 'distance', as well as develop a knowledge base that would enable a degree of empathy with practitioners, I endeavoured to learn taxidermy.

My 'learning' of taxidermy began with a one and half-day course with a local taxidermist after which I practiced on my own, spending hours bent over a desk in a Scout hut in the town of Aberystwyth, documenting my progress using film and jotting notes in a research diary. This produced some interesting comparative comments on the different moles I was working with and their materiality, for example, their size, the damage done by the trap that killed them, their levels of decomposition, how I worked with these issues, and my (slowly) improving skill. I found, however, that the camera footage did not parallel the liveliness of my field notes, which detailed tentative dealings with decomposition and frustrations with slicing through skin. As Simpson has highlighted, "while video can capture the minute detail of practices and allow for the close analysis of this, video in and of itself does not necessarily present or give a sense of the affective relations present in such encounters" (2011: p.343).

During this process of solo practice, filming and methodological reflection, I discovered that others around me were keenly interested in the process of taxidermy. I proposed to show them what I knew under the proviso that I could film our sessions and interview them about their experience afterwards. I found these sociable practice sessions were fruitful in providing more eventful film footage. They inserted a participatory aspect into the methodology in as much as the researcher and researched created the empirics together, exploring how best to work with dead non-human bodies and talking over our reactions to and about the process. Consider, for example, the following transcript that documents a moment from a taxidermy 'learning' session where, due to a shortage of mole bodies, fellow learner Dawn and I took turns it in turns to taxidermy one mole. At the stage relayed below, I work on the mole while Dawn takes photographs.

ME: Ok this one must have started rotting before they froze it.
DAWN: Oh really?

[Once the incision is made I work quite quickly, prizing the carcass from the skin as it has started decomposing. The carcass seems to have less hold on the skin and falls away fairly easily. I cut through the two back legs and after 10 minutes am now ready to pull the skin over the top of the carcass.]

ME: Oh my god Dawn!
DAWN: Is it ming?

[I pull skin back over the back.]

ME: This is disgusting. It's all rotting.
DAWN: Oh my word, that isn't anything like the ones we did. [Dawn takes some photos over my shoulder.] Oh fucking hell, I just got a whiff of that. [Dawn walks out of the shot.] Stinky!
ME: Oh my god! [I dust some chalk over the skin and get back to skinning.]
DAWN: Go on you can do it, you can do it!
ME: At least it is making me work quite quickly.
DAWN: Yeah.
ME: It's just, it's really hard, there is like, hardly any definition really, and it's hard to get the tissue tension to cut off right, so I am cutting bits of muscle off when normally it is really easy not to.

(24/08/2010)

The decaying mole body encountered here, and the tactile struggle that ensued, signals that temporality is not only a facet of living bodies, but also dead ones. Temporality is a key proponent of Irigaray's thinking on the body, for example, for it highlights the changeability or 'amorph' of corporeal materiality and its "ceaseless reshaping of incarnation" (Ziarek, 1998: p.63). Irigaray makes this argument through recourse to the material movements of her feminine body, where "mucous membranes evade my mastery" (1993: p.170) as they move from inside to outside and change her materiality. This temporal understanding of the body highlights that it is forever in a "state of imperfection … [a result of] the unfinished condition of every living being" (Irigaray, 1993: p.193).

Engaging in the process of 'learning' to do taxidermy with other humans present generated discussion on process, and highlighted body language and actions. It drew attention to senses other than touch, as well as outlining the context and evolution of events as they happened. I began to consider a different ethics, that of "[d]ifference-in-relation" (Whatmore, 2002: p.159). This is an ethics, Harrison states, which reveals a, "concern with ways of living, folding and becoming with 'more-than-human' others" (2006: p.132). In my practice and interviews, as I outline below, I began to also consider the effect of dead animals on the living human taxidermist.

Experiencing skin: learning taxidermy

Taxidermy is a monotonous and slow process. Hands move rhythmically over the skin and flesh of an animal; from the initial incision into the skin with a scalpel, to the prising of skin from flesh, to the use of tweezers and scissors in the removal of tissue and fat from the skin's underside once separated from the body. The repetitive acts of pulling skin taught to create enough tension in the tissue connecting skin to carcass so that you can easily slice through it – then pull and slice again, pull and slice – create a rhythm that appears to have an 'enduring predictability' (Edensor and Holloway, 2008).

The monotony created by such rhythms enables the practitioner or, as I discuss in this section, taxidermy learner, to "get sucked into the task [because] you are so focused on such a tiny little area, just with the scalpel" (Dawn, pers. com. 27/01/2010). Yet, as Edensor remarks, referencing Lefebvre "there is no identical absolute repetition indefinitely…there is always something new and unforeseen that introduces itself into the repetitive" (2010: p.6). In this section, I draw attention to those moments in which repetitive, rhythmic actions are thwarted by unforeseen movements that occur as a result of pressure applied to one area of a carcass, taking effect in another area of a carcass. These are unforeseen results that take visceral effect on the living body of the taxidermy learner.

As Dawn's comment intimates, the taxidermy learner can get 'sucked into the task' of removing skin from an animal's body; as a result, the rest of the relations between its various unseen component parts (for example, bones or muscle in contact with organs) can be forgotten. As fellow learner Dawn skinned her first mole, she gripped onto the shoulders and head of the animal to hold it down as she pulled the skin over towards its nose. Suddenly she felt movement between her fingers as the head popped out of its socket. This was a release of tension that not only resulted in the movement of bone beneath the mole's flesh – a very different sensation to the passive, monotony that had characterised the process of skinning so far – it also travelled into Dawn herself: she jumped in her chair with surprise. After the taxidermy session I questioned her about the episode. In response she said:

> I don't find it physically repulsing because I have done veterinary work and stuff before, it's just not a nice feeling of squeezing something and having it pop … almost more like a reflex than a fear.
>
> (pers. com. 27/01/2010)

Dawn's embodied reaction was one that emerged in response to an aarthymic juncture in her repetitive engagement with the mole body on the table. It occurred during an encounter with a different, internal corporeal space. This arrythmic moment was materialised in movement, a movement inadvertently set in motion by Dawn attempting to manipulate and manoeuvre the dead non-human material. This experience highlights the reciprocity of the sense of touch in that the dead animal body was moved into action by Dawn's hands, while she herself was also in turn affected by the 'feel' of that action. Reciprocity is central to Irigaray's concept of porosity where living bodies can touch and be touched, can feel and be felt, signalling the non-hierarchal character of this sense. For Irigaray, bodies touching become open to the effects of one another; she explains "[t]he internal and external horizons of my skin interpenetrating with yours wears away their edges, their limits, their solidarity. Creating another space – outside my framework. An opening of openness" (1993: p.59). In the case of taxidermy, this reciprocity is asymmetrical as the dead can no longer 'feel'; however, as Dawn's response demonstrates, the sense of touch can still instigate response in both bodies.

Such arrhythmic moments of reciprocal action do not only occur when hidden, internal body spaces are pressured into action; they can also occur between the external spaces of carcass flesh and skin. Consider the following film transcript in which, a fellow learner and I attempt to skin our first squirrel.

ME: Ow, how far do you have to go with the tail? [I ask in full knowledge that I will not get an informed answer.]
ROGER: All the way! [Roger offers a suggestion.]
ME: *Sigh.* [I keep pulling and slicing, pulling and slicing through tissue.]

[Roger sits next to me working on a front leg, he nears the toes.]

ROGER: Aw that's gross. You pull and the skin just peels off.

ME: I'm not getting anywhere with this bloody tail. I have got cramp in my left hand [I pick it up and dangle the squirrel in front of me.] *Sigh.*

ROGER: I guess yours is still happy that he has got a tail! [In his own frustration with the tail Roger cut the tail off so that he could continue taking the skin over the back towards the head.]

ME: hmm?

ROGER: I guess yours is probably still happy he has got a tail!

ME: em.

[We continue to work in silence. I am slicing the tissue between tail carcass and tail skin, then pulling the skin down towards the end of the tail to enable access to more tissue to be sliced. Suddenly the tension of the tissue connecting skin to carcass is less than that created by my pulling action and my hand whips away from the carcass and is left clutching the skin and fur.]

ME: err! Oh my god! [I sit back in my chair away from the body in front of me.]

[Roger looks up from what he is doing and laughs loudly.]

ROGER: You got the tail out!

ME: That is so gross! [I laugh for some time. I pick up the tail carcass and explore the bone joints.] I guess you can just pull it out then!

(23/03/2010)

The sudden change in movement, and the halt of tension proprioceptively deployed and felt by the muscles in my arm, initiated a visceral affect. My reaction was somewhat different to Dawn's as, rather than jump, my sense of shock manifested through nervous laughter. Shock here emerged from a lack of 'somatic reflexivity' (Paterson, 2015), an undeveloped understanding of the degrees of sensation encountered during the skinning process. Paterson explains that "[s]uch reflexivity is premised on the proprioceptive ability to 'feel' and therefore recognize one's own body and its movements" (2015: p.48). This lack of recognition, resulting in the surprise, shock, and for me ensuing laughter, signals the emergence of a new relation between myself and the dead squirrel matter as I learn what degree of tension holds the carcass together, and thus how much pressure is needed to pull sections from one another.

While the process of working on a body may for the most part be repetitive and monotonous, wherein a sense of familiarity can develop with each stroke of the scalpel, or pluck of the tweezers, a carcass' materiality and its occasional outbursts of liveliness highlight the absence of the animal in life through the shock of re-animation. However, this absence of life is not associated with the life of the other, but the slippage of matter, a kind of 'thing power' that enables the dead bodies to "manifest traces of independence or aliveness, constituting the outsideness of our own experience" (Bennet, 2010: p.xvi). In such instances, the practitioner develops an understanding of the animal being worked on and a 'located' subjectivity develops from the mingling of subject and matter. For experienced taxidermist Ben, this is something that is "fundamentally difficult to rationalise". He expands, responding to a comment I made on the potential agency of an animal's dead body:

> ... you could almost watch it get killed on the road outside and think, this is going to be absolutely fine and it's almost as if the thing itself, it doesn't want to play, it doesn't want to be preserved or whatever. Because I know, I know for a fact that there is nothing wrong with it. I know from a technical level that there is absolutely nothing wrong with it, but the damned thing falls apart and I've had things that have been in freezers for twenty five years, literally and every part about them has said, this will not be any good, er, but you persevere with it and it turns out better than a fresh one. It's almost as if there is something um, er, built within it, some inborn personality if you like that has decided that it doesn't want to be thrown away, and you could never rationalise that could you.
>
> (pers. com. 05/10/2009)

Ben's comments resonate here with Ruddick's suggestion that there are "an infinite array of possibilities of becomings that can only emerge in practice" (2010: p.41), for it is only in opening the freezer and practicing taxidermy on the animal stored that Ben develops his understanding of dead animal potentiality. Ben's comments echo the struggles faced by myself and Dawn in the process of learning this practice, wherein the materiality of a carcass can alter the 'feel' of what is being worked on; here we find that the internal and external become entwined through "the ability of strange substances to cross the subject's own boundaries and in so doing, change the very contours of identity" (Fullagar, 2001: p.180). Ben takes this a step further, to suggest the vibrancy of a carcass can have the ability to deflect and alter intended creative direction.

Through the effects created by the dead vibrant matter, which acts against Ben's aims for taxidermy mount presentation, the bodies appear to impose their own agency upon his creative attempts at re-animation. Such objects "can become vibrant things with a certain effectivity of their own... a liveliness intrinsic to the materiality of the thing formally known as an object" (Bennett, 2010: p.xvi). Dead animals may well be conceptualised as a 'social body' in a "constant state of de- and re-composition in relation to other bodies" (Ruddick, 2010: p.28). Bodies, dead and alive, emerge into and out of various material relations to become entangled in the production of material affects. In the next section, I turn to focus on the body of taxidermists, to flesh out other ways in which bodies become entangled in the making taxidermy.

Transferral of tension: the connectivity of matter

While the previous section highlighted the tensions, frictions and textures that, in the main, occurred between the bodies of the taxidermy learner and dead animal, this section attends to the work of more experienced taxidermists and the role that human emotions can have upon the dead matter being worked on. Acknowledgment of this capacity further complicates the notion of a distinct inside and outside in taxidermy practice. Considering the connectivity between flesh and self, dead matter and taxidermy practitioner, I then turn to consider ethical imperatives involved in taxidermy practice.

Working on an animal to produce a taxidermy specimen is to always work with an absence, the sense of lost life. In mimicking life, intimate attention is paid to the anatomy of an individual animal and painstaking work is carried out to produce a new body. What is to be created, if done 'well', is an echo of the form that used to be, that of the individual animal in life. However, this form will always be an unfamiliar version, taking on as it does certain elements and experiences of its creator's skill, techniques, style and even emotion. The latter is something that the taxidermist needs to consider as the following comment made by Jack highlights.

> I have to be relaxed and comfortable in my surroundings in order to produce good work. If not the tension in my body travels down my arms into whatever it is I am creating.
>
> (pers. com. 07/30/2010)

This transferral of tension from body to body highlights the porosity between taxidermist and animal, a transferral that occurs when the skin of each meets and acts as the medium through which the soma take effect. In the interview extract that follows, Jasmine, a London-based artist, draws out the need for staying calm when conducting taxidermy. After having returned to London from two weeks of volunteer work under Jack's instruction, Jasmine told me the following:

> ... the one thing that I learnt [pause] the most from him was how to be calm and watching him and also how he has regular tea breaks. Because before I was doing something and waiting until I had done it and then having tea break and by the end of that I would be so frustrated and I would either rip apart the thing I was doing or just make it look rubbish... [now] I think I have got a lot more control and I understand exactly what is going on. And so, today, I have never been able to skin a blue tit, and today I have managed to skin one whole and I didn't make one hole in it and that's since coming back. And I play classical music and yeah, it all about being patient and calm.
>
> (pers. com. 15/06/2010)

What these extracts suggest is the vulnerability of a dead animal's form to the subjectivity of the taxidermist. In working with the body of an 'other', Jack and Jasmine render the specimens as not only vulnerable in terms of physical contact, but also in terms of their emotive capacities. This understanding positions vulnerability as "a form of sociality in which there is a world in common" (Fullagar, 2001: p.180); and which thus "generates an ethical moment in which the flesh and mortality of the self and other is felt". If touch, as Irigaray argues, creates a form of sociality then it is pertinent here to consider the ethics involved in moments of contact through a framework that takes the social into account.

Such a framework has been considered by Beasley and Bacchi (2007), who aim to draw out an understanding of care ethics that digresses from neo-liberal and paternal projectionist framings. Instead they argue for an approach termed 'social flesh': a political metaphor for "thinking critically about politics, interconnection and sociality" (2007: p.291). While this vocabulary focuses on human embodied interdependence, I would suggest that a vital materialist approach could implement this perspective to 'things' if we follow Irigaray's argument for the sociality of touch and Braidotti's recognition that objects are also subjects of our inquiry. As both matter and life are capable of touch and being touched, both catalysts for affect, then we can consider both as interconnected through a form of sociality. However, there is an issue at stake for individual responsibility if such an approach is taken. As Bennett states:

> In emphasising the ensemble nature of action and interconnections between persons and things, a theory of vibrant matter presents itself as simply incapable of bearing *full* responsibility for their effects.
>
> (2010: p.37, emphasis in original)

In response to this problematic, Bennett states that "[p]erhaps the ethical responsibility of an individual human now resides in one's response to the assemblages in which one finds oneself participating" (ibid.). Certainly, taxidermy practitioners, in compliance with various laws as well as their own ethical philosophies, illustrate a variety of responsible responses to working with dead matter.

In modern day commercial taxidermy, the disclosure of legal and ethical documentation goes hand in hand with the careful monitoring of exotic and endangered species as a means to control the unlawful killing of animals.[1] The Guild of Taxidermy, an institution set up in the UK to assist the profession, recommends the use of and supplies log sheets for the taxidermist to record the details of the animal to be worked on. For example, if an animal was found by a passer-by on the road, details of who found it and where, as well as information on where it was kept before the taxidermist was given it, date of death, sex and age (if known), must be logged.[2] Furthermore if any animal registered under Annex A is sold in a commercial manner, appropriate article 10 exemption certificates must be obtained and kept with the mount for legal purposes. Responsible practice is, therefore, enforced by law and upheld by practitioners keen to maintain the legal and ethical validity of their profession.

Conclusion

Geographic discussions of making have drawn our attention to the practices, materials and skills that bring humans and non-humans, both organic and inorganic – e.g. stone, glass, wool, wood – together in various ways. In so doing, geographers have argued that making facilitates connections between human makers (Price, 2015), such that we might understand making as a process that brings people in touch. In this chapter, I have offered an in-depth ethnographic account of working with dead animal bodies in the practice of making taxidermy. In so doing I have paid close attention to the touching of both dead non-human and living human bodies, the physical connections through which tensions, frictions and pressures overflow from one body to the other. Unpacking the micro-encounters (Carr and Gibson, 2016) of taxidermy in this chapter, I have highlighted the agency of a non-human animal carcass within an assemblage of making. I suggest that paying attention to the micro-encounters of taxidermy practice – moments of connections between human and non-human – enables a democratic approach to the making process, as it gives space to consider the role and impact of 'earth others' (Bradotti, 2013).

Primarily, taxidermy practice inverts our usual experience of skin. Whereas in most activities we encounter its surface presentation – that is, the top layer of the epidermis – in taxidermy time and effort is spent with the under layers. Taxidermy presents skin as something to be pulled or torn, actions that can not only further its fragility or durability, depending on the animal, but also set in motion a vibrancy of skin, tissue and carcass. In this chapter, I have considered the arrhythmics that occur in taxidermy practice to highlight such moments and in turn looked to the manner in which, dead non-human animal matter becomes lively within the human and non-human assemblage. Focusing on the haptic quality of such moments has drawn attention to the development of proximate knowledge in learning to undertake taxidermy. For example, such knowledge develops with regards to the strength and tensions of tissue, skin and bone, and how these respond to human action. Tensions and pressures, I have illustrated, flow from live body to dead matter (and back again), such that a taxidermy mount emerges through the interconnectivity of both live and dead bodies: through vibrant matter.

Using the sense of touch to account for the vibrancy of dead matter in an assemblage has highlighted the interconnection between bodies, and goes some way to producing knowledge that reduces the centrality of the sovereign human as a knowing subject in the research process. Attending to micro-intensities felt by the taxidermy practitioner and produced by the body of an animal in an assemblage, whilst then being sensitive to the effects of the same, provides a means of accounting for the singularity of an animal's body. To be sure, the human is very much still present within the making process here; however, looking to the sense of touch provides a way to account for the reciprocity of tactile encounters. Touch allows us to produce knowledge *with* non-human animals, be they dead or alive.

Notes

1 As outlined by the Wildlife and Countryside Act 1981 Part one section 6 and 11.
2 See log sheet provided by Mike Gadd of the Guild of Taxidermists at http://www.ta xidermy.co.uk/uk/taxidermy-law/taxidermy-log-sheet/ [Accessed 23/08/2015].

References

Beasley, C. and Bacchi, C. (2007) Envisaging a new politics for an ethical future: beyond trust, care and generosity – towards an ethic of 'social flesh'. *Feminist Theory*, 8(3), 29–298.

Bennett, J. (2010) *Vibrant Matter: A Political Ecology of Things.* Durham: Duke University Press.

Braidotti, R. (2010) The politics of 'life itself' and new ways of dying. In Coole, D.H. and Frost, S. (eds) *New Materialisms: Ontology, Agency, and Politics.* Durham: Duke University Press, 201–220.

Braidotti, R. (2013) *The Posthuman.* Cambridge: Polity Press.

Brown, M. (2013[1884]) *Practical Taxidermy.* Worcestershire: Read Book Ltd.

Carr, C. and Gibson, C. (2016) Geographies of making: rethinking materials and skills for volatile futures. *Progress in Human Geography*, 40(3), 297–315.

Davies, G.F. (2006) The sacred and the profane: biotechnology, rationality and public debate. *Environment and Planning A*, 38(3), 423–444.

Davies, G.F. (2003) Editorial: a geography of monsters? *Geoforum*, 34(4), 409–412.

Dixon, D.P. (2008) The blade and the claw: science, art and the lab-borne monster. *Social and Cultural Geography*, 9(6), 671–692.

Dixon, D.P. (2009) Creating the semi-living: on politics, aesthetics and the more-than-human. *Transactions of the Institute of British Geographers*, 34(4), 411–425.

Dixon, D.P. and Straughan, E.R. (2010) Geographies of touch/touched by geography. *Geography Compass*, 4(5), 449–459.

England, K. (1994) Getting personal: reflexivity, positionality, and feminist research. *The Professional Geographer*, 46(1), 80–89.

Edensor, T. (2010) *Geographies of Rhythm: Nature, Place, Mobilities and Bodies.* Surrey: Ashgate.

Edensor, T. and Holloway, J. (2008) Rhythmanalysing the coach tour: The Ring of Kerry, Ireland. *Transactions of the Institute of British Geographers*, 33(4), 483–501.

Foucault, M. (2003 [1973]) *Birth of the Clinic.* London: Routledge.

Fullagar, S. (2001) Encountering otherness: embodied affect in Alphonso Lingis' travel writing. *Tourist Studies*, 1(2), 171–183.

Harrison, P. (2006) Poststructural theories. In Stuart, C.A. and Valentine, G. (eds) *Approaches to Human Geography.* London: Sage.

Haraway, D.J. (2008) *When Species Meet.* Minneapolis: University of Minnesota Press.

Harper, D. (2017) Taxidermy. *Online Etymology Dictionary.* Available from: http:// www.etymonline.com/index.php?term=taxidermy [Accessed 26th July 2017].

Hawkins, H. and Straughan, E.R. (2014) Nano-art, dynamic matter and the sight/sound of touch. *Geoforum*, 51, 130–139.

Hodgetts, T. and Lorimer, J. (2015) Methodologies for animals' geographies: cultures, communication and genomics. *Cultural Geographies*, 22(2), 1–11.

Irigaray, L. (1993) *An Ethics of Sexual Difference*. London: Cornell University Press.

Lorimer, J. and Driessen, C. (2013) Bovine biopolitics and the promise of monsters in the rewilding of Heck cattle. *Geoforum*, 48, 249–259.

Ruddick, S. (2010) The politics of affect: Spinoza in the work of Negri and Deleuze. *Theory Culture and Society*, 27(4), 21–45.

Shildrick, M. (2001) Some speculations on matters of touch. *Journal of Medicine and Philosophy*, 26, 387–404.

Simpson, P. (2011) 'So, as you can see...': some reflections on the utility of video methodologies in the study of embodied practices. *Area*, 43(3), 343–352.

Straughan, E.R. (2015) Entangled corporeality: taxidermy practice and the vibrancy of dead matter. *GeoHumanities*, 1(2), 363–377.

Patchett, M. (2008) Tracking tigers: recovering the embodied practices of taxidermy. *Historical Geography*, 36, 17–39.

Patchett, M. (2012) On necro-ornithologies. *Antennae: The Journal of Nature in Visual Culture*, 20, 9–26.

Patchett, M. (2015) Witnessing craft: employing video ethnography to attend to the more-than-human craft practices of taxidermy. In Bates, C. (ed.) *Video Methods: Social Science Research in Motion*. London: Routledge, 71–94.

Patchett, M. and Foster, K. (2008) Repair work: surfacing the geographies of dead animals. *Museum and Society*, 6(2), 98–122.

Paterson, M. (2015) On aisthesis, 'inner touch' and the aesthetics of the moving body. In Hawkins, H. and Straughan, E.R. (eds) *Geographical Aesthetics: Imagining Space, Staging Encounter*. Surrey: Ashgate, 35–54.

Paterson, M. (2007) *The Senses of Touch: Haptics, Affects and Technologies*. Oxford: Berg.

Paterson, M. (2009) Haptic geographies: ethnography, haptic knowledges and sensuous dispositions. *Progress in Human Geography*, 33(6), 766–788.

Polanyi, M. (2009[1966]) *The Tacit Dimension*. Chicago: The University of Chicago Press.

Price, L. (2015) Knitting and the city. *Geography Compass*, 9(2), 81–95.

Whatmore, S. (2002) *Hybrid Geographies: Natures, Cultures, Spaces*. London: Sage.

Wildlife and Countryside Act (1981) Available from: http://www.legislation.gov.uk/ukpga/1981/69 [Accessed 8th September 2012].

Wylie, J.W. (2005) A single day's walking: narrating self and landscape on the Southwest Coast Path. *Transactions of the Institute of British Geographers*, 30(2), 234–247.

Wylie, J.W. (2006) Depths and folds: on landscape and the gazing subject. *Environment and Planning D Society and Space*, 24(4), 519–535.

Ziarek, E.P. (1998) Toward a radical female imaginary: temporality and embodiment in Irigaray's ethics. *Diacritics*, 28(1), 60–75.

10 Knitting the atmosphere

Creative entanglements with climate change

Miriam Burke

The five women of the 'knit and natter' group settle into chairs in a horseshoe shape. It is a warm summer evening, and in the small classroom the windows and doors are open and a cool breeze blows in through the trees outside. Maggie is still standing and offers to make Richard a cup of tea, poking in the cupboards in the small kitchen area at the back of the room. There's lots of cooing and giggling as the ladies in the group take in their surroundings, admiring the children's drawing of plants and animals, a few taxidermy wildfowl and a little 'nature scene' created from felt and pipe-cleaners. We are in the East Reservoir centre of the London Wildlife Trust (LWT), and the aptly named Richard Van Neste is preparing to talk to the group about the impacts of climate change on the local area.

Maggie, an enthusiastic 65 year old, well-known and well-liked in the local area for her voluntary work with elderly people and the local community centre, begins the conversation by enquiring about bats in the local area, and the conversation begins at a roll into tales of childhood and fears of bats getting tangled in hair... Richard appears a bit perplexed by the direction of the conversation and reassures the six women and girls present that bats are nothing to be afraid of. He gently returns the topic of conversation to climate change, what we had come to the centre to talk about...

RICHARD: One of the first concerns is how wildlife affects us. This is a perfect time of year and a perfect moment to talk about that, because the last few weeks have been really hot, haven't they?

GROUP: affirmative mumbling.

RICHARD: It's been so hot the last few weeks. And we've seen in the last 10 or 15 years some record breaking heat waves. We haven't seen the record set this year for record temperature, highest temperature ever recorded in this country, was in Faversham in Kent, so not all that far away, an hour and a half from here, and it was 38.5 degrees.

MARY: Ooh wow! 35? That's as hot as Africa!

RICHARD: yeah

MIRIAM: 38!

MAGGIE: When was that recorded?

RICHARD: That was 2003 I believe.

RICHARD: We had big heatwaves in 2003 and 2006. In 2003, throughout the whole of northwest Europe, over a period of two weeks we had a really big heatwave. I was 13 years old.

OLIVE: Chuckles

RICHARD: I bet if you go back and look at holiday snaps and things, you'll be able to picture it. But 20,000 people died during that heatwave. That they could attribute to the heatwave. They didn't die "of heatwave" but they died of heart attacks, strokes, breathing difficulties.

BETTY: We [older people] get more tired in this weather.

RICHARD: They probably had underlying conditions, but it shortened their life, so it was 20,000 people above what they would have expected. Hospitals were inundated with people; real, real problems ...The conversation expands and progresses from painting train-tracks white to cope with increased temperatures, housing construction in African villages to keep inhabitants cool, grass roofs, little Egrets, magpies, crows and blackbirds, to problems with pigeons and squirrels entering Mary and Betty's homes. Tea is drunk, and biscuits shared. Some of the women have brought their knitting with them, and Olive offers to help Richard cast-on a row – although instead, the group opts for a walk around the 'private' area of the nature reserve on the reservoir as this is the part that they do not usually get to see. We head out onto the East Reservoir as the sun is setting over the London skyline and enjoy being outside in a secluded urban wildspace in the warm summer evening.

Introduction I: the group

Building on my practice as an artist, this empirical research is from a participatory art project, which set out explicitly to engage people with the idea of climate change. This chapter pays attention to practices of making, and the ways they offer novel insights into embodied and emotional relationships with other people, with nonhumans and the wider environment, with particular reference to the climate. The fieldwork was carried out on the Woodberry Down estate, in North London between May 2014 and May 2015. In order to set the scene, I will first describe the set up of the group and the project before moving on to the theoretical analysis of its making practices.

The PACT knit n natter group meets every Monday evening in the Redmond Community Centre, in Woodberry Down. The PACT knit n natter group was set up as part of the PACT[1] (Prepare, Adapt, Connect, Thrive) project run by social enterprise Manor House Development Trust. PACT was a £1 million, three year project funded by the Big Lottery to promote sustainability and raise awareness of climate change in Woodberry Down which ended in 2016. The project had many strands, including waste reduction, reducing energy consumption, training local people for 'green' employment, gardening and cooking projects. The knitting group was initially set up in 2013 and funded by the project as a way to bring people together to knit draught excluders for doors. The group has received other funding from another Big Lottery stream (Well London), and as of the time of writing, continues independently from the (completed) PACT project.

The knitting group is made up of a core group of five women who come every week, myself included; a further 12 women and children who drop in regularly, plus people who come either as a one off or either occasionally. For the purpose of this study, I will concentrate on the core members of the group. The women, Maggie, Betty, Olive, and Mary are all long-term residents of the estate who either live in the council properties or newly built social housing. Maggie and Olive are pensioners and while Betty and Mary work, all are in receipt of some form of income related benefit. They have little prior knowledge about climate change,[2] but most are actively involved in other local community activities through the community centre, demonstrating a level of enthusiasm and care for their local area. The group has done a number of other collective projects including knitting blankets for the elderly, hats to raise money for a homeless charity and poppies for Remembrance Day.

The knitting group meets in a community centre on the edge of Hackney's 'East Reservoir' nature reserve, managed by the London Wildlife Trust. The initial trip described at the beginning of this chapter was designed to introduce the knitters to their immediate 'wild space' as well as the ways the area and its inhabitants are changing as the climate warms. The group chose to make a collaborative artwork that would represent what climate change means to them and their local area. The group is particularly concerned about the impact of rising temperatures in the summer. And so, collectively, we decided to knit a parasol to celebrate the role that their local greenspaces play in keeping urban areas cool. A knitted parasol, strung over the carcass of an old umbrella for structure is populated with knitted representations of the plants and animals that live in the nearby nature reserve: leaves, mushrooms, flowers, a family of mice, an oversized toad, a grey slug.

Introduction II: my theoretical lens

Climate change, and the age of the anthropocene, calls us to recognise the influence we have on the natural environment. It is clear that humankind has irreversibly changed the planet we all inhabit and this has been described as a global experiment, one in which we all play a part (Gibson-Graham, 2011). Although it was never a very accurate description of life in the biosphere, climate change makes all the more apparent the way that we have separated ourselves from nature, highlighting the false distinction between humans and nonhumans, economy and ecology, thinking and acting, since even our most mundane actions have an impact on the global climate (Dalby, 2013; Gibson-Graham, 2011). In light of the reciprocal relationship that we have with our environment – critics of separateness and separation thinking are asking us to think connection, rather than separation, interdependence rather than autonomy – as a way to think about our place in a changing world (Gibson-Graham, 2011). Indeed, many scholars would go so far as to say that if the current ecological crisis is not to be a prelude to despair, then it must be an incitement to improvisation, experimentation and risk taking (Clark, 2014; Gibson-Graham, 2011; Haraway, 2016). Inspired by Gibson-Graham's (2011) call for a spirit of connectivity, experimentation and active participation, this chapter considers the role of making practices as a way to not only think about the environment and environmental change, but also think with the environment. I examine the ways that this kind of participation has the potential to lead to proactively ignite new ways of considering ourselves as intimately connected to our wider environment through the material act of knitting.

For Gibson-Graham (2011), an 'experimental orientation is another way of making (transformative) connections; it is a willingness to "take in" the world in the act of learning, to be receptive in a way that is constitutive of a new learner-world, just as Latour's concept of "learning to be affected" describes the formation of new body-worlds' (Latour, 2004; Roelvink and Gibson-Graham, 2009). In experimentation there's no active transformative subject "learning about" a separate inert object, but a subject-object that is a "becoming world"' (Gibson-Graham, 2011: p.8). This experimentation also involves a spirit of participation, and in this research my own position is one of both researcher and participant. I recognise that we are all participants in the 'becoming world' and I embrace my part in this.

This chapter is primarily about connectivity. It considers how, in the process of knitting a parasol the knitters create a network of connections to their environment that did not exist before; and how this connectivity is transmitted to others. First, I start by reviewing the need to 'bring climate change home' and the ways in which scholars have suggested this could be done. This leads me to suggest that, since making practices are well known for being sources of connection, the practice of making, and knitting in particular, offers an opportunity for creative, embodied connection to the environment. To focus this argument, I look at the way that knitting is simultaneously an imaginative practice and a physical one. I begin by looking at the idea of knitting as a collective imaginative practice generally before thinking specifically about the role of charisma in the knitter's imaginative decision making process; and finally how the practice of knitting connects the knitters to their nonhuman neighbours. The following section draws on how Gibson-Graham (2011) suggests that Jane Bennett's (2010) ideas on the vitality of matter can serve as a way to not simply see connection, but to understand connection in a physical way, through the process of knitting. Finally, I consider how the knitted creatures have become sites of connection; through the processes described in this chapter they have gained an ability to transmit connection.

Connection and the climate: why bringing 'climate change home' matters

Although it is arguably one of the most pressing concerns of our time, most people in the west are, at best, ambivalent towards climate change (Hulme, 2009; Lorenzoni et al., 2007; Ockwell, Whitmarsh & O'Neill, 2009; Poortinga et al., 2011). Information about climate change is often presented as a series of abstract, and sometimes contradictory sets of scientific statistics and data, with an emphasis on the physical causes and impacts of climate change, which tend to spread over space and time in complex relationships (Duxbury, 2010). Many commentators note that for many people, climate change is difficult to comprehend or connect with in an appreciable way; leading individuals in western societies to think that the impacts of climate change are spatially and temporally distant to their own lives (Duxbury, 2010; Hulme, 2009; Slocum, 2004).

As such, many scholars are calling for ways to 'bring climate change home' (generally understood to be a western, urban home) and examine how – as a concept – it takes on meaning on an everyday level (Brace and Geoghegan, 2010; Ereaut and Segnit, 2006; Hulme, 2011). Climate change may be a global issue, but it is played out within everyday scales, and a focus on smaller scales of meaning may hold the key to understanding the ways in which climate change is relevant and matters to people's everyday lives. Global issues have an impact on an everyday scale, because issues are deeply rooted in time and space specific contexts (Gibson-Graham, 2002; Massey, 2005).

In order to attend to these small scale and localised understandings of climate change, this chapter draws on the work of feminist scholars who argue that a focus on the everyday and embodied experience is of crucial importance to understanding larger scale political worlds. It is within these mundane and often domestic interactions that large scale political relations are produced, reproduced and maintained and yet, they also contain within them the possibilities for altering, undermining and undoing these relations (Dyck, 2005; Mitchell et al., 2004). Therefore, paying heed to the ways in which people are already connecting to, and engaging with nonhumans, materials and environmental changes can potentially afford crucial understandings on which to build as ecological and economic pressures increase.

It is in these small scale, domestic settings that we get a specific and located understanding of what an issue as vast as climate change means to people in their everyday lives. Knitting is a practice that has (until fairly recently) often been disregarded and overlooked because it is considered mundane and domestic (Price, 2015). But herein also lies its value; knitting is both an affordable and accessible hobby and a skilled making practice. Therefore, it has the potential to enable participants to express feelings and emotions about a large topic, such as climate change, through mundane, domestic and familiar processes. This research builds on burgeoning research into how making practices have the potential to connect humans and nonhumans, particularly in regards to environmental change (de Leeuw and Hawkins, 2017; Hawkins et al., 2015; Ingram, 2013).

Connection through knitting: making is connecting

Maggie is the first of the group to get to the community centre. While she waits for the kettle to boil in the kitchen, she clears leaflets from a table and begins to rearrange the chairs. I arrive to find her exasperated, pulling at the arm of a sofa to move it to a more suitable position. I help her move the sofa. Together we arrange cups, teabags, milk and biscuits onto two trays and carry them to the circle of chairs, along with a large, insulated canister of just boiled water – one of those plastic silver ones with a large black button on the top that is always somehow awkward to push. I retrieve the bags of yarn, needles and patterns from the cupboard in the corridor. Familiar faces begin to arrive, Betty offers to make Olive, Maggie and me, a cup of tea. She leaves Maggie's tea black; she knows Maggie is very particular when it comes to the ratio of sugar, milk and tea and does not want to risk getting it wrong. Mary is not here yet. Betty rummages in her bag to see if she has received a message as to whether Mary is coming. No message, so Betty texts Mary to ask if everything is ok. Olive, Maggie and I sit down and we begin to knit.

In essence, knitting is the creation of a fabric from a single thread, formed into horizontal rows of individual loops that intermesh with each successive row of loops (Black, 2012: p.4). Knitting is a labour intensive activity; the craft requires skill, muscle memory and embodied knowledge of yarn, in order to produce knitted fabric. Knitted items are often made and given as gifts, and in the process of making, the items are imbued with affection and care (Turney, 2013). Studies have demonstrated the powerful emotional connections between knitter, the practice of knitting and knitted material; indeed, the practice of knitting is a way to make and to maintain emotional worlds for those who practice it. (Black, 2012; Corkhill et al., 2014; Gauntlett, 2011; Price, 2015; Rosner and Ryokai, 2009; Shin and Ha, 2011).

Cups of tea, text messages, welcoming hugs; these things are what keep the knitters coming to the centre each week as much as the knitting. The knitter's friendships are based first and foremost on a shared enjoyment of knitting; the women did not know each other before the formation of the group, but they now class one another as friends. It is well documented that social relationships are forged and maintained through the regular meeting of the participants and in the repetitive actions of knitting together; and the PACT knit and natter is no exception (Collier, 2011; Corkhill et al., 2014; Kenning, 2015).

Knitting as a craft has moved from a traditional and often solitary practice to a contemporary movement, which encompasses knitting as a practice that is performed and displayed in public with the development of yarn bombing, knitted protests and large online communities (Price, 2015). Communal knitting practices, particularly the activity of yarn bombing, which involves placing knitted items in public spaces, have become actively politicised in recent years; studies have critiqued how knitting is put to work to embody powerful political forces, feminist activism and contributes to both the shaping and subversion of gender roles (Bratich and Brush, 2011; Mould, 2014; Myzelev, 2009; Pentney, 2008; Price, 2015; Turney, 2013). It is in these communal and collective practices that important connections and social relationships are made. There are connections between yarns and needles; materials and emotions; and between participants of the group, leading Gauntlett (2011) to note that making is connecting (see also Ingold, 2013).

During the sessions, knitting techniques are taught, shared and discussed within the group. The social relations of the group consist of mundane inter-actions (such as making tea) as well as the knitting specific interactions; the group supports the knitter's making practice and one another's emotional needs as one and the same. In this way, the knitted objects that the group produce are more than the work of one pair of hands. Entwined with the loops and the yarn, the social ties and the communal creativity are an intrinsic part of the construction of the knitted items.

Knitting as an imaginative practice

The members of the knitting group are all highly skilled and use a huge amount of creativity in their craft. Yet, when I ask them about their skill and creativity level, they are incredibly modest – both in regards to their skills in making and their skills in imagination and idea generation. "Oh, I can do a few things, but I'm nowhere near as good as Beulah", says Maggie, while knitting an intricately patterned baby blanket, perhaps the third one of its kind this year. Olive just chuckles when I ask her about how creative she is; "oh, I don't have creative bone in my body dear. No, arty things just aren't for me, although I do have a few watercolours up at home that I like." The other members of the group reflect this sentiment that they view themselves as resolutely uncreative. For them, knitting is something they do and they love – and while I see it as very much an expression of skill and creativity – the knitters themselves do not see it in this light. For them, it is something familiar, easily accessible and sociable – indeed it is the very "old-fashioned-ness" of knitting that they enjoy. It offers these women an ordinary craft; creativity with specific boundaries, within which they feel safe and secure, and because of these boundaries they have the opportunity to express themselves without feeling threatened.

And yet, the group is embroiled in a highly creative collective decision making process in deciding how best to visually represent climate change. The topic that hovers in the group the longest is about which creature they are choosing to knit. The choice of creature is usually an individual decision, influenced by the yarn and the patterns to which the knitter has access, although the final decision is most often made privately at home during the week, rather than at the two hourly meeting of the group itself. The knitters take this task seriously, they think about their own experience of the wetlands, the creatures that they themselves have seen, or that they know to live there. They hold these creatures in mind as they flick through books magazines and loose leaves of knitting patterns. It is a local, specific and creative decision making process about their local experience and the ways that climate change may affect their environment.

Mainstream imagery of climate change has often focussed on melting icebergs, polar bears, scientific graphs, drought and starvation; and as a result, visual representations of climate change have often been critiqued as being clichéd, losing emotional impact and even turning people away from the whole issue altogether (Dixon, 2009; Doyle, 2007; O'Neill et al., 2013; O'Neill and Nicholson-Cole, 2009; Sheppard, 2012; Yusoff, 2010; Yusoff and Gabrys, 2011). Such a strong focus on often barely imaginable, future impacts and the effects on specific species has also served to obscure the importance of biodiversity and the interdependence of humans and nonhumans (Yusoff, 201; Yusoff and Gabrys, 20110). By focussing on attempting to picture climate change as a global issue, many large scale media "climate change communication" campaigns have also effectively ignored, and undermined, the role of local understandings and connection to the environment.

The knitting group's representation of climate change is an interesting departure from much mainstream imagery because the knitters have chosen such unlikely creatures to represent climate change. Unlike polar bears, a knitted slug (along with earthworms, mushrooms and toads) do not have the kind of charismatic appeal we usually associate with large mammals, but they obviously and intriguingly, have a certain appeal to the knitters. What influences the creativity and the imagination that is enabled through the practice of knitting? Why have these women chosen to knit a toad, a slug or an earthworm over a squirrel, an egret or a tree? The knitters had access to patterns for all of these creatures (or at least ones that could easily be altered), but many of the creatures they were most keen to knit were those that appear at first glance, 'uncharismatic', such as a grey slug.

Knitting as imaginative practice: decision making and the role of charisma

The ladies coo and giggle when they compare what animals they have knitted over the week. Slug, knitted by Olive brought on some fits of laughter as Maggie, Betty and Devon tried to work out where its body ended and its head began. They all agreed that it was not biologically correct, but all three knitters squeezed it and caressed it with affection. Maggie thought it looked more like a rabbit than a slug, she was a little confused that its eyes did not appear to be in the correct place, and tried to manipulate its shape by bending it at the 'neck' to make it more 'slug-like', although it resolutely popped back into the same position as soon as she let go.

Figure 10.1 Grey slug. May 2015
Photograph by Miriam Burke

Lorimer (2007) outlines different types of charisma: ecological and aesthetic among them. Ecological charisma is shared across cultures and describes the way that different organisms orientate and differentiate themselves. This concept of charisma is based on the idea that humans are a certain size and shape and we rely predominantly on our (spectrum-specific) sense of vision; nonhumans whom we can easily relate to, say those that are of a similar size and shape to us and who also rely heavily on vision are thought of as having more 'ecological charisma' than others. So, for example, a monkey, a bear or a bird are more likely to fall into this category than say a microbe or a rock. Aesthetic charisma is where human affections are triggered by appearance. Human beings have an instinctive preference towards creatures that look like our own babies, so of those beings that have ecological charisma, those that have disproportionately large heads and eyes tend to be even more appealing (Milton, 2002). Animals such as otters or kittens fall neatly into this category.

The knitters have clearly still chosen to knit things that have ecological charisma, they have chosen things that they can see, can hold, and can attend to, from flowers to mushrooms to invertebrates and a handful of small mammals. But they have rejected those with the most aesthetic charisma – for instance hedgehogs or the many diverse migratory birds that over winter on the reservoirs embody if not strictly aesthetic charisma, a certain romanticism; yet these are not the creatures chosen by the group. Instead, perhaps there is an element of 'feral charisma' working here, a charisma grounded in a sense of respect for the other and its wildness (Milton, 2002). The knitters comment on the fact that the items they are knitting are unusual, they are not the things most retired women knit; there is something about this mild rebellion that touches the knitters, they giggle about confused comments from their husbands. Here, there is an importance placed on the representation of something that is different or unusual; at its core, it is this otherness and wildness that appeals to them. Yet, in this wildness, there is also familiarity and it is in this liminal state between comfort and discomfort, familiarity and the unknown that seems to be most appealing for the choice of creature. The knitters are thinking about the creatures, as they knit, they are 'holding them in mind' and by way of this unusual charisma and thought process they are connecting to the real creatures, living in and around the reservoirs.

Knitting as imaginative practice: as a way to connect to nonhumans

Maggie has knitted a family of four mouse babies and a mother mouse. Alone in her flat, she thinks of the group and wonders what the other's will think.

Figure 10.2 Family of pink mice. May 2015
Photograph by Miriam Burke

She thinks about the pet mice she carried to school in her pockets as a child, and her mind wanders to other, less appealing mice, who made a home in her old flat with its damp corners and disembodied nightly noises. And the mice on the reservoirs, brown, wild-eyed and unassuming. She wonders how they are coping in the light evening rain.

The knitted creatures are a way of connecting the knitters emotionally to the world around them; these creatures are not the ones living in far off lands, affected by distant and often unimaginable forces, but the knitters' local, and unusually charismatic, neighbours. Animals have traditionally been the main way of humans understanding environmental change through their migrations and movements and in this way animals extend our sensory capacity to both notice and respond to climate change (Yusoff, 2010). Through their practice, the knitters are extending an emotional relation to the world and, in doing so, are bringing the idea of climate change home, to their very doorstep.

Perhaps, these choices are also influenced by the local city assemblage in which knitters and critters reside. The dark waters of the reservoir loom outside the community centre's windows and these have seeped into the imagination of the knitters, the creatures and critters scurrying and living their unknowable lives metres from their doorstep have played a part in the process of detailing and imagining the living relations and entangled futures with the nonhuman neighbours. These are the imaginative and creative emotional connections sustained through the process of knitting. Knitting is also an essentially physical practice, and in its very physicality it can also offer other forms of connection, based on materials as well as imagination.

Knitting as a physical practice: a way to connect to materials

Maggie's mother mouse is about half the size of the palm of her hand, knitted from soft pink, acrylic wool that has been donated to the group from an elderly lady elsewhere on the estate who cannot physically attend the knitting group because of her poor health. It is stuffed with the remnants of an old, stuffed toy that had been ripped apart by a dog, which Betty had brought to share with the groups a couple of weeks previously. The eyes are made from beads from an old cushion that has been picked apart for its constituent parts.

The materials the knitters use have mostly been re-appropriated from other things. Very little is wasted, and this is something that is of utmost importance to the knitters. Although the group has a small pot of funding – enough at least to buy some yarn, needles and keep the group in tea and biscuits – the members prefer to bring in their own items, and reuse things that they already own, or that are being thrown out. There is a strong ethos of not wasting materials. Although the women say that they are not doing this for environmental reasons, there are clear environmental benefits to this 'make do and mend' attitude to creating their knitting.

The yarn they tend to use is cheap, readily available acrylic yarn, it has the feel of static electricity as it runs through fingers on its way around and about knitting needles, and the fabric produced is reminiscent of mass-market clothing. Somewhat ironically in making work about climate change, the manmade yarn is itself a by-product of the oil industry; its constituent parts are made from the bodies of prehistoric sea creatures, which swam in warm, tropical seas when the world was much warmer. This yarn is teased out, knotted and looped, it is moved by hands and warmed by skin. In handling the material of the yarn, the knitters are making embodied connections; their muscles learning new ways to manipulate yarn and create fabric; the yarn manipulating muscles in a reciprocal relationship between knitter and knitting (Ingold, 2013). The material make-up of the yarn becomes a physical, material connection to the earth and to the specific history of climate change.

The knitters make clear, distinct choices about which materials to use; and this choice of materials reflects different modes of connection to the knitter's local environment. In the 'feminist project for belonging in the anthropocene', Gibson-Graham (2011) draws on the work of Jane Bennett (2010) to consider how we can go beyond simply seeing connection, to a situation where we are actively and physically connecting with the more than human. They argue, after Bennett (2010), that in essence, we are all just different collections of the same stuff, bacteria, heavy metals, atoms, matter-energy. By thinking about materiality in this way – as 'vibrant', the connection of human and nonhuman is not only about affection and imagination, but also about a shared, material identity where we are all part and parcel of the same stuff (Gibson-Graham 2011: p.6). So, in the working of the yarn and the practice of making, the knitters are not only engaging and connecting with the idea of climate change, they are, in fact, connecting with the very stuff and material processes of climate change.

The yarn is moved and manipulated, it is wrapped, twisted and looped, and small creatures are formed. But these connections to the environment exist not only for the knitters, since, through the handling and sharing of the knitted items, these connections are also transmitted and communicated to others through the material bodies of the creatures.

Knitted items as sites of connection

The six women sit in a circle on sofas and hard plastic chairs, and bring in items they have made over the previous week to show the other group members. They handle one another's knitting, turn it over, examine the stitching and comment on the amount of stuffing; they chat about what they have made and what they like or dislike about the creatures they have created. Some people bring questions: how best to manufacture eyes or noses, or how to rescue a half worked stitch pattern hovering precariously across four needles.

The knitters hold one another's knitting; they pass around their creations, in a similar way they pass around my four week old baby when I bring my daughter to meet them for the first time. They react to the creatures, turn them over, peer at them through varifocal lenses and hold them at arms' length to get an idea of them at a distance. They place them on one another's shoulders and chuckle. They speak about their own experiences of making while cradling each other's knitting. The process of sharing knowledge, opinions and affection is as much about the physical sharing of items as it is about the emotional or intellectual understandings. In their knitted forms, the creatures take on a dimension of aesthetic charisma that the original creatures arguably do not have. Through the process of knitting, the perception of the slug, for instance, has changed from a seemingly cold, slimy creature to a fuzzy, cosy one. It has taken on new kinds of charisma, by way of the physical and emotional process of knitting.

The embodied, emotional reactions to the creatures are a mix of visual, tactile and olfactory qualities to the very fabric of the creatures (Pajaczkowska, 2005). But it is not only the touch of the fabric that has specific, material resonances; Barnett (1999: p.28) argues that to hold or wrap a body in fabric is a prosthesis of touch, it is "the continued existence of the hand in the absence of the body that offers touch". In this sense, the knitters are not only touching the knitted creatures, but they are also touching – and connecting to – one another through the prosthesis of the knitting. The action of hands extends from hand to yarn, and from yarn to hand, between material and human reciprocally. In her recent book, *Staying with the Trouble*, Donna Haraway (2016) suggests that a community of crocheters who have gathered to make a crocheted coral reef have stitched "intimacy without proximity", a presence without disturbing the critters that animate the project. Intimacy without proximity is not 'virtual presence' it is 'real presence', but in loopy materialities. Like the crotched coral reef, the parasol "is a practice of caring without the neediness of touching by camera or hand in yet another voyage of discovery" (Haraway, 2016: p.79).

Summary: making is connecting to the climate

Gibson-Graham (2011) describe two, interrelated but distinct forms of connection to the environment; the first is like belonging to the world as one does to a family, suggesting an affect of love and an ethic of care. Over this chapter I have sought to demonstrate how, through their practice, the knitters have connected emotionally to one another, and have extended this care to the nonhuman creatures that inhabit their local environments. I have done this by looking at the idea of the role of charisma in forming emotional bonds with creatures, and thought about the ways in which the knitters develop affection for the things they knit.

By attending to the materials used in the knitting project, I suggest that a different kind of connection to the local environment is also made, one that stems from the vibrancy of the yarn itself, and the ways that the materials have influenced the group and their thought processes. This, perhaps comes close to Gibson-Graham's second project of connection, which involves belonging within a "heterogeneous monism of vibrant bodies" (Bennett, 2010: p.119) which they describe as "attuning ourselves more closely to the powers, capacities and dynamism of the more than human", and recognising ourselves as part and parcel of the world; we are nature and cannot be separated from it (Gibson-Graham 2011: p.7). Finally, I have suggested that the making, passing and handling of items can be seen as a way to transmit and communicate these new networks of connections.

Over the year that the knitting group has gathered to chat, drink tea and create knitted things, the knitters have also gathered around the topic of climate change – a topic they have not given much thought to before joining the group. By way of their making practices, they have supported and engaged with one another and engaged in creative processes that have connected them in new ways to the creatures who live on the reservoirs; developing new understandings for how they, as city dwellers, are intimately (inter)connected to the environment and to one another. Through the emotional and material aspects of the practices of making, there is a sense that the knitters have indeed gathered of new ways of thinking connection.

Notes

1 http://www.mhdt.org.uk/pact/
2 Asked about the causes of climate change in a focus group, replies of group members included air pollution, cars, litter and volcanoes. When asked what they could do about it, the group focused on ways to reduce landfill waste by recycling and composting food waste.

References

Barnett, P. (1999) Folds, fragments, surfaces: towards a poetics of cloth. In *Textures of Memory: The Poetics of Cloth*. Nottingham: Angel Row Gallery.

Bennett, J. (2010) *Vibrant Matter: A Political Ecology of Things*. Durham: Duke University Press.

Black, S. (2012) *Knitting: Fashion, Industry, Craft*. London: V&A Publishing.

Brace, C. and Geoghegan, H. (2010) Human geographies of climate change: landscape, temporality and lay knowledges. *Progress in Human Geography*, 35, 284–302.

Bratich, Z.J. and Brush, M.H. (2011) Fabricating activism: craft-work, popular culture, gender. *Utopian Studies*, 22, 234–260.

Clark, N. (2014) Geo-politics and the disaster of the anthropocene. *The Sociological Review*, 62, 19–37.

Collier, A.F. (2011) The well-being of women who create with textiles: implications for art therapy. *Art Therapy*, 28, 104–112.

Corkhill, B., Hemmings, J., Maddock, A. and Riley, J. (2014) Knitting and well-being. *Textile*, 12, 34–57.

Dalby, S. (2013) The geopolitics of climate change. *Political Geography*, 37, 38–47.

De Leeuw, S. and Hawkins, H. (2017) Critical geographies and geography's creative re/turn: poetics and practices for new disciplinary spaces. *Gender, Place and Culture*.

Dixon, D. (2009) Creating the semi-living: on politics, aesthetics and the more-than-human. *Transactions of the Institute of British Geographers*, 34, 411–425.

Doyle, J. (2007) Picturing the clima(c)tic: Greenpeace and the representational politics of climate change communication. *Science as Culture*, 16, 129–150.

Duxbury, L. (2010) A change in the climate: new interpretations and perceptions of climate change through artistic interventions and representations. *Weather, Climate and Society*, 2, 294–299.

Dyck, I. (2005) Feminist geography, the 'everyday', and local-global relations: hidden spaces of place making. Susanne Mackenzie Memorial Lecture. *The Canadian Geographer/Le Géographe canadien*, 49, 233–243.

Ereaut, G. and Segnit, N. (2006) *Warm Words, How We Are Telling the Climate Story and Can We Tell it Better?* London: Institute for Public Policy Research.

Gauntlett, D. (2011) *Making is Connecting: The Social Meaning of Creativity from DIY and Knitting to YouTube and Web 2.0.* London: Polity Press.

Gibson-Graham, J.K. (2002) Beyond global vs. local: economic politics outside the binary frame. In Herod, A. and Wright, M. (eds) *Geographies of Power: Placing Scale*. Oxford: Blackwell.

Gibson-Graham, J.K. (2011) A feminist project of belonging for the anthropocene. *Gender, Place and Culture*, 18, 1–21.

Haraway, D. (2016) *Staying with the Trouble: Making Kin with the Chthulucene.* Durham: Duke University Press.

Hawkins, H., Marston, S., Ingram, M. and Straughan, E. (2015) The art of socio-ecological transformation. *Annals of the Institution of American Geographers*, 105, 331–341.

Hulme, M. (2009) *Why We Disagree About Climate Change: Understanding Controversy, Inaction and Opportunity.* Cambridge: Cambridge University Press.

Hulme, M. (2011) Meet the humanities. *Nature Climate Change*, 1, 177–179.

Ingold, T. (2013) *Making: Anthropology, Archaeology, Art and Architecture.* London: Routledge.

Ingram, M. (2013) Washing urban water: diplomacy in environmental art in the Bronx, New York City. *Gender, Place and Culture*, 21, 105–122.

Kenning, G. (2015) "Fiddling with threads": craft-based textile activities and positive well-being. *Textile*, 13, 50–65.

Latour, B. (2004) How to talk about the body? The normative dimension of science studies. *Body and Society*, 10, 205–229.

Lorenzoni, I., Nicholson-Cole, S. and Whitmarsh, L. (2007) Barriers perceived to engaging the public with climate change among the UK public and their policy implications. *Global Environmental Change*, 17, 445–459.

Lorimer, J. (2007) Nonhuman charisma. *Environment and Planning D*, 25, 911–932.

Massey, D. (2005) *For Space.* London: Sage.

Milton, K. (2002) *Loving Nature: Towards an Ecology of Emotion.* London: Routledge.

Mitchell, K., Marston, S. and Katz, C. (2004) *Life's Work: Geographies of Social Reproduction.* Oxford: Blackwell.

Mould, O. (2014) Tactical urbanism: the new vernacular of the creative city. *Geography Compass*, 8, 529–539.

Myzelev, A. (2009) Whip your hobby into shape: knitting, feminism and the construction of gender. *Textile*, 7, 148–163.

O'Neill, S., Boykoff, M., Niermeyer, S. and Day, S.A. (2013) On the use of imagery for climate change engagement. *Global Environmental Change*, 23, 413–421.

O'Neill, S. and Nicholson-Cole, S. (2009) "Fear won't do it": promoting positive engagement with climate change through visual and iconic representations. *Science Communication*, 30, 355–379.

Ockwell, D., Whitmarsh, L. and O'Neill, S. (2009) Reorienting climate change communication for effective mitigation: forcing people to be green or fostering grass-roots engagement? *Science Communication*, 30, 305–327.

Pajaczkowska, C. (2005) On stuff and nonsense: the complexity of cloth. *Textile*, 3, 220–249.

Pentney, B.A. (2008) Feminism, activism, knitting: art the fibre arts a viable mode for feminist political action. *Thirdscape: A Journal of Feminist Theory and Culture*, 8, Available from: http://journals.sfu.ca/thirdspace/index.php/journal/article/view Article/pentney/2010 [Accessed 3rd May 2017].

Poortinga, W., Spence, A., Whitmarsh, L., Capstick, S. and Pidgeon, N. (2011) Uncertain climate: an investigation into public scepticism about anthropogenic climate change. *Global Environmental Change*, 31, 1015–1102.

Price, L. (2015) Knitting and the city. *Geography Compass*, 9, 81–95.

Roelvink, G. and Gibson-Graham, J.K. (2009) A postcapitalist politics of dwelling: ecological humanities and community economies in conversation. *Australian Humanities Review*, 46, 145–158.

Rosner, D. and Ryokai, K. (2009) Reflections of craft: probing the creative process of everyday knitters. In *Proceedings of the Seventh ACM Conference on Creativity and Cognition*, ACM. 195–204.

Sheppard, S. (2012) *Visualising Climate Change: A Guide to Visual Communication of Climate Change and Developing Local Solutions*. Abingdon: Earthscan.

Shin, H.Y. and Ha, J.S.H. (2011) Knitting practice in Korea: a geography of everyday experiences. *Asian Culture and History*, 3, 105.

Slocum, R. (2004) Polar bears and energy-efficient light bulbs: strategies to bring climate change home. *Environment and Planning D: Society and Space*, 22, 1–26.

Turney, J. (2013) Making love with needles: knitted objects as signs of love? *Textile*, 10, 302–311.

Yusoff, K. (2010) Biopolitical economies and political aesthetics of climate change. *Theory, Culture and Society*, 27, 73–99.

Yusoff, K. and Gabrys, J. (2011) Climate change and the imagination. *Wiley Interdisciplinary Reviews: Climate Change*, 2.

11 A sustainable future in the making?

The maker movement, the maker-habitus and sustainability

Rebecca Collins

"That's the thing about makers – if they see something that's broken, they want to fix it."

(Ken)

Introduction

Recent years have seen the emergence of what has been termed a new 'maker movement' (Anderson, 2012; Hatch, 2013). Sometimes cast as an essentially new mode of engagement with the practices and potentialities of making instigated by new (digital) technologies (e.g. Hatch, 2013), other times as a (re)turn to the fundamentals and rewards of traditional crafts (e.g. Hackney, 2013), and by yet others as a combination of these (e.g. Gauntlett, 2011; Lindtner, 2015; Luckman, 2013), opportunities to practice making in an array of forms are increasingly widespread. Whilst the maker movement may be viewed as a mere continuation of the motivations, sensibilities and knowledges of established hobbyist or artisanal making practices, it may equally herald a shift in how 'making' is perceived and practised in a global consumer society. At a time when human appropriation of, and impact on, the material world has led to the defining of a geological epoch characterised by our material traces, there are important questions to be answered about how we can fulfil our social, cultural and psychological needs as human beings through relationships with material things that leave fewer negative environmental impacts.

In this chapter I explore how the maker movement might constitute a valuable mechanism through which the emergence of more environmentally sustainable material cultures in consumer economies might be facilitated. After a brief introduction to the movement and its relationship to both established making practices and contemporary consumer economies, I articulate a conceptual 'maker-habitus' – an embodied orientation to the material world characterised by an interest in material (re)production. I argue that fundamental to the 'maker-habitus' is a particularly acute affordance sensitivity – that is, an ability to identify the potentialities of materials and material things. Recent empirical work with a diverse range of 'makers' is used to illustrate this notion at work and, in turn, to suggest that increasing societal support for the proliferation of such sensibilities might be key to eliciting a more environmentally sustainable everyday material culture.

The maker movement and consumer culture

Two key features characterise what has come to be known as the 'maker movement'. First, its constitutive spatialities are particularly diverse. Whilst household sheds, garages, spare rooms and dining tables remain key sites of hobbyist making practice, in the twenty-first century digital technologies have helped to transform the distribution, scale and scope of amateur making. The practices once hidden in the garden shed are now, via the internet, connected with those of makers worldwide. Websites such as instructables. com, as well as the growing repository of "How do I...?" tutorials that exist on YouTube, allow globally dispersed amateur makers to share, learn and problem-solve beyond the traditional boundaries of the domestic sphere. This 'global workshop' has been identified as fundamental not only to the sharing of skills and enthusiasm that perpetuate, evolve and grow hobbyist practices (e.g. Craggs, Geoghegan & Neate, 2013; Fox, 2014; Gauntlett, 2011; Geoghegan, 2013; Sennett, 2009), but also to the emergence of "micro revolts" (von Busch, 2010) against the status quo – such as the homogenisation and 'black-boxing' of the majority of today's consumer goods (Dant, 2010).

Secondly, and in a sense a result of this spatial patterning, it has been argued that the maker movement democratises production (Anderson, 2012). Whilst it should be acknowledged that this democratisation is, at best, partial,[1] the shift of production processes back into the hands of everyday people, facilitated by digital technologies – as predicted 35 years ago by Alvin Toffler in *The Third Wave* (1980) – seems to be coming to pass. Indeed, far from being "anti-modern" (Campbell, 2005), craft – in the form of artisanal production and hobbyist making, such as sewing, knitting and woodwork – has re-emerged[2] as a networked collection of socially, culturally and politically relevant practices, producing material things seen – at least by those with particular kinds of economic and cultural capital (Luckman, 2013) – as preferable to objects of mass production (Campbell, 2005, citing Kopytoff, 1986; cf. Lindtner, 2015). As Campbell (2005: p.28) identified over a decade ago, "one of the intriguing features of modern consumer society is the way in which machines have become reappropriated by the craft tradition, aiding and abetting craft consumers rather than robbing them of their traditional autonomy".

The mainstreaming of alternative or countercultural values that this might be seen to represent has been described by Castells (2012) as reflective of the strain under which global capitalism is now placed. Indeed, Comor (2010) argues that the humanising and empowering capabilities of 'prosumption' (when items are both produced and consumed by/for the same person; see also Jackson, 2010; Ritzer, 2014) have the potential to strike a significant blow to the alienation that characterises contemporary consumer society.[3] By re-situating the alienated consumer within production-consumption processes where their autonomous action directly shapes the material world they inhabit, small-scale making practices can be viewed as aligned with the defetishised, environmentally aware, socially co-operative and personally fulfilling consumption required by more sustainable lifestyles (Brook, 2012; Crawford, 2009).

In sum, the increasingly networked global community of making enthusiasts that comprise the maker movement emphasises the intrinsic human desire for connection (Gauntlett, 2011) as well as throwing down a challenge to passive consumer culture (Lindtner, 2015). In a sense this is about feeling part of a community – being able to learn and gain esteem through the recognition of skill or creativity. However, it is equally about how first-hand interaction with the material world, through manipulation of its composite parts, can engender a more visible and visceral understanding of our impact upon it (Jackson, 2010; Krzywoszynska, 2015; Price, 2015). It is with this context in mind that the concept of maker-habitus is elaborated here.

Defining the maker-habitus

The notion of maker-habitus describes the idea that we might be better able to perceive creative (re)uses for material things if we have had the opportunity to build up a bank of embodied experiences with a range of materials, tools and techniques. The more opportunities we have to employ these skills in new contexts, the better able we are to identify creative solutions to new problems.

Maker-habitus is thus conceptualised as a means of explaining how persistent engagement with making practice (Figure 11.1, left) might, through cumulative embodied experience and the resultant sense of competence (effectivity), engender inclination towards the repair, reuse and repurposing of material things (Figure 11.1, right). It describes an orientation to the material world characterised by a capacity to perceive the affordances (Dant, 2005; Gibson, 1986; Ingold, 1992) of material things – that is, what they offer by way of potential uses in the environment in which they are situated. Implicit within this is an ability to transpose learning and action from one context to another, in the course of which embodied experiential knowledge may be developed or refined. The concept of maker-habitus draws on ideas from three

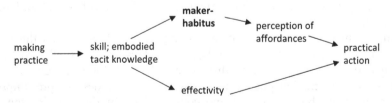

Figure 11.1 Maker-habitus genesis and effect
Diagram by Rebecca Collins

theoretical strands: Pierre Bourdieu's notion of habitus; ideas about tacit, embodied knowledge discussed in the work of (amongst others) Michael Polanyi and Maurice Merleau-Ponty; and the notion of affordance, discussed first by James Gibson in the 1960s and elaborated more recently by Tim Ingold and Tim Dant.

Despite its focus on the situated and contextual development of orientations to the physical-social world, Pierre Bourdieu's concept of habitus has been somewhat underused by geographers (although note Bauder, 2005; Pimlott-Wilson, 2011). Habitus refers to the dispositions, norms or tendencies into which an individual is (implicitly or explicitly) socialised, and which, in turn, guide subsequent thought and action (Bourdieu, 1984; Wacquant, 2005). It is produced through interactions as individual agency responds to structures encountered in the course of everyday life, and the learning accrued through these interactions becomes embodied and unconsciously reproduced in subsequent scenarios (Bourdieu, 1984). Traditionally habitus has been used to explore issues related to social class and social capital, such as experiences of education and employment (e.g. Bauder, 2005; Cushion & Jones, 2014; McDonough, 2006; Smyth & Banks, 2012), precisely because of its capacity to make sense of the transferability of dispositions between contexts, as well as between individuals and even different generations. However, there is, equally, scope for its application in contexts where understanding of dispositions oriented at the material world is sought (see, for example, O'Connor, 2007), particularly in light of Bourdieu's interest in cultural capital. Although his interests in capital and habitus were not developed as part of the self-same theoretical project, they are inextricably connected, the social and cultural 'landscape' within which dispositions are formed necessarily shaping the specific characteristics and manifestations of the habitus.

In sum, the notion of habitus as an experientially produced, embodied and transferable disposition to the socio-material world forms the first conceptual building block for the 'maker-habitus'. The 'maker-' prefix is developed in response to two further theoretical strands. The first pertains to the genesis of tacit, embodied knowledge which subsequently informs responses to stimuli. Michael Polanyi, in his work on the transmission of skill, argues that repeated effort to internalise new knowledge or skill brings about a psychological change resulting in a form of embodied knowing capable of expression only through action (Gamble, 2001, with reference to Polanyi, 1958). Maurice Merleau-Ponty, in *The Phenomenology of Perception* (1981), similarly concerned with the connection between environment and embodied knowledge, suggests that, "our gaze, prompted by the experience of our own body, will discover in all other 'objects' the miracle of expression" (p.197), hinting at the capacity for embodied knowledge to result in 'seeing' the possibilities that exist within surrounding 'objects'. Thus both Merleau-Ponty and Polanyi point to the emergence of a potentially transformative relationship between human subject and material objects as a result of regular, direct interaction.

Recent scholarship concerned with the tacit, embodied knowledge that characterises making has demonstrated empirically how accumulation of such knowledge creates greater sensitivity to the potentialities of materials as a result of deep haptic human–subject/material–object 'entanglements' (e.g. Straughan, 2015 on taxidermy; Paton, 2013 on stone carving). Erin O'Connor's work on glass blowing has, similarly, shown how the cumulative acquisition of embodied knowledge produces a capacity to anticipate, to read or feel through (human–)material–tool interactions what action is required in a given moment (O'Connor, 2007; see also Eden & Bear, 2011). Additionally, the identification of these moments or processes of 'entanglement' as uniquely individual, personal and both spatially and temporally situated – and thus, a function of the specific environment in which the practice is situated – frames acts of making as an intersection of multiple material micro-geographies, dispersed across and drawing from various times and spaces (Ingold, 2000; Patchett, 2016). The fundamental role of the environment in which making practice is situated informs the second strand of the 'maker-' prefix through reference to concept of 'affordances'.

First articulated by psychologist James Gibson[4] in the context of his work in visual perception and ecological psychology (1986), and subsequently developed by anthropologist Tim Ingold (e.g. 1992) and sociologist Tim Dant (2005), 'affordances' are ascribed to objects depending on how they are understood in a particular context. "[D]epending on the kind of activity in which we are engaged, we will be attuned to picking up a particular kind of information, leading to the perception of a particular affordance" (Ingold, 1992: p.46). This concept is fundamental to the articulation of maker-habitus, since it is through the process of perceiving an affordance and acting to bring that affordance to fruition that maker-habitus is manifested in practice. Ingold (1992) describes the specific form of agency required to enact affordances in terms of effectivity – the possession of specific competencies necessary for the act of transformation. Since competence can only be achieved through practice, first-hand manual interaction with the material world is fundamental to the genesis and manifestation of maker-habitus (see also Dant, 2005; Frow, 2003; Ingold, 2013).

By drawing together these theoretical threads concerned with disposition, embodiment and perception, the suggestion is that through experience of the materialities and potentialities of a wide range of material things – via practices of making, mending, maintaining, tinkering, taking apart – it is possible to become particularly attuned to alternative forms those things could take. Thus, the concept of maker-habitus seeks to unite a focus on the development and deployment of knowledge in the socio-material world with human sensitivity towards the affordances of material things within that world, and the E/environmental[5] implications thereof. It goes beyond the ineffable transmission of skill, as it equally seeks to articulate how tacit, embodied knowledge, gathered through first-hand experience, creates and informs an orientation to the world oriented around the (re)production of material things.

The value of a theorisation with this function has been made clear by scholars from a wide range of disciplines concerned with our actions within and towards the material world. Richard Sennett (2009) in *The Craftsman*, for instance, suggests that, "we become particularly interested in the things we can change" (p.120), and that we therefore need to understand what we can change, and how. Isis Brook (2012) has argued that the transformative potential of acting on material things to bring them back into use provides satisfaction and affirms one's productive competence. She suggests that "active, purposive engagement with the material realm" can help us to "reintegrate ourselves into the material fabric of the world" in ways attuned to environmentally sensitive consumption (p.109). Further, empirical studies focused on object repair and maintenance (e.g. Cooper, 2005; Gregson, Metcalfe & Crewe, 2009; Maller, Horne & Dalton, 2012) have shown that people are more likely to take care of objects that they have had some involvement in the (re)production of. Indeed, Graham & Thrift (2007: p.2) refer to the "emancipatory potential" of engaging in acts of reproduction, maintenance or repair and, through those practices, gaining new knowledge of the material world.

Underpinning several of these studies is a concern with tackling the wastes of consumer culture, which evidence suggests troubles more consumers that we might think.[6] A maker-habitus orientation to the world, by virtue of its preoccupation with the potential of materials, may help attune consumers to the 'liveliness' or 'vibrancy' of matter (Bennett, 2010; Straughan, 2015), even when on the cusp of disposal or wastage. Whilst Jane Bennett (2010: p.14) argues that we need to "cultivate the ability to discern non-human vitality", the vitality of matter can only do so much without the addition of human agency to direct and influence the manifestation of that vitality, and give that manifestation meaning.

These brief examples suggest potential for us to play a more visible part in shaping our immediate material environment and express agency within a consumer-material culture in which, despite the pernicious myth of 'choice', it is increasingly constrained. Acknowledgement of making as a co-productive act (e.g. Ingold, 2013; Paton, 2013) allows the maker to feel greater ownership over their material possessions (see, for example, Jackson, 2010; Torrone, 2005; Yarwood & Shaw, 2010), and brings them closer to the material world through their participation in shaping it. In doing so the broader human impact on shaping the world through how we act with its material constituents may, itself, be brought a little closer.

In the discussion that follows, I draw on conversations with 15 amateur 'makers' of varying kinds in order to illustrate the genesis, embodiment and potential of the 'maker-habitus'. This is followed by some reflections on what these makers' experiences suggest about the theoretical and practical utility of the concept for encouraging more environmentally sustainable modes of interaction with the material world.

What makes a maker?

The 15 participants whose experiences inform this discussion were a diverse group. They ranged in age from 21 to early 70s, with at least one participant per age decile. Most participants were in their 30s or 40s. Five were female and ten male. A wide range of employment backgrounds were represented, from student to organic farmer, retired teacher to IT consultant. Four of the participants were employed in professions which may be viewed as particularly closely aligned with their hobbyist making: Mike was a builder-carpenter; Bob was an electrician; Sig designed electronic components; and Barbara was a retired textile technology teacher.[7]

The participants were recruited and interviewed in two phases. Eleven participants were members of a makerspace in a university town in the east of England. A makerspace is a community-run communal space dedicated to a wide range of making practices. They usually house a range of the latest digital making technologies including laser cutters, 3D printers and CNC milling machines, as well as hand-held tools such as drills, planes, screwdrivers, etc. Some also accommodate textile, ceramic and glass craft activities by providing equipment such as sewing machines and kilns. Ethnographic work was conducted with the makerspace group over a period of 18 months in 2013–2014. The four additional participants were based in or close to a second university town in the north-west of England. They volunteered their skills as part of a university-run sustainability event focused on waste avoidance and repair, and were approached for interviews following the event. Where possible, the interviews took place in a location associated with making practice, such as the makerspace. In cases where this was not possible, interviews took place in local coffee shops or the participant's workplace.

Before discussing these participants' experiences, it is worth reflecting momentarily on how their diversity reflects what was described earlier in terms of a (re)emergent maker movement. Those participants allied with the makerspace could be seen as epitomising the 'new' maker movement. The majority sought out that space because of their interest in – and desire to use – the technologies it possesses, particularly items such as the laser cutter. Yet as the discussion below reveals, they came to that space with an interest in making based on exposure to a wide range of 'traditional' making practices. These 'traditional' practices are also represented by the four participants recruited from the university 'repair fair'. The nature of the making they practised had no immediate need – superficially at least – for digital technologies or a shared working space. Yet they, too, straddled the divide with the 'new' maker movement through their desire to connect both digitally (through websites) and through communal, social and sharing-based interactions. Together, this group of participants reflects a rounded picture of the purported 'maker-movement' – a fundamentally hybrid entity of traditional and digital making practices, skills and motivations.

Over the course of the discussions with the 15 participants, three dominant facilitators of the genesis of a maker-habitus emerged. These can be summarised in terms of 'play', 'place' and 'people'.

Play

There was common reference amongst the respondents to the importance of fun and playfulness – both in childhood and in adulthood – in developing fascination with materials and their capabilities. All participants spoke about the extent to which they were encouraged (or not) as children to experiment with materials and tools, and all those who were parents spoke passionately about the value they placed on providing their own children with that creative freedom. The enduring impact into adulthood of parentally sanctioned play activities in childhood has been vividly evidenced in Yarwood & Shaw's (2010) work with model railway enthusiasts, many of whom referred to parents' encouragement of their childhood modelling activities as underpinning their adult hobbyist practice. Barbara and Nina both spoke about the expectation placed upon them by their parents that they would find creative ways of entertaining themselves as children. Barbara (early 70s, retired textile technology teacher) reported how she was not allowed to sit "doing nothing" as a child, and that if she wanted to read she had to knit at the same time. Nina (mid 30s, university researcher) suggested that the limited resources she had for play forced her to be creative. For Bob (early 40s, electrician), this link between acknowledgement and empowerment drove his commitment to bringing his two young sons to the 'family'[8] sessions at the makerspace: "so they realise they can make shelves, rather than buy them".

The theme of learning by 'breaking' further demonstrated the importance of play. Mike (early 40s, builder-carpenter), like Bob, learned most of his making skills through "a lot of messing about [laughs] and breaking things!" and now encouraged his two children to do the same. This penchant for breakage was, in essence, about playfulness – referring to learning through processes of 'breakage' simply seemed more legitimate than 'playing'. Bob noted the social unacceptability of adults 'playing', saying, "I think it's a positive thing to reinforce skills and learn new skills. So I think it's all a part of the... human thing of needing to create and make and use tools. [...] People look at playing as wasted time." Yet the vast majority of the participants readily admitted the joy they found in 'playing'. Mike, for instance, said, "I love whimsy... and nonsense... and to be somewhere where people will spend days on end struggling to make some piece of nonsense work, is brilliant." Here, Mike's self-confessed penchant for whimsy connects with Mann's (2015) recent consideration of the same phenomenon in the context of urban yarn-bombing (see also Price, 2015). Mann's suggestion that whimsy can increase attentiveness to what has become habitual and mundane in our (material) environment underlines the value of playfulness in drawing critical attention to material configurations that might benefit from remaking of some kind.

Simon (late 40s, IT consultant) enthused about making a "project for a project's sake [...] even if it's just a little robot that beeps". Such enthusiasm for a "pointless machine" (Simon) is arguably indicative of a strong sense of self-efficacy produced through simultaneous (and mutually generative) enjoyment and skill in making practice. Whilst there is a potential tension here related to the unsustainability of making "pointless machines", I acknowledge these experiences because of their importance in explaining the role of playfulness in developing a love of making and, thus, a maker-habitus. Further, and as elaborated by Mitchell (2011) with reference to practices of 'modding'[9], whilst such 'pointless' practices are often driven by the pursuit of fun, they exist in the same time-space as other material practices simultaneously framed as both modding and repurposing/remaking. I return to the implications of this tension in the chapter's conclusion.

Place

In their ethnographic exploration of resourcefulness in the home, Wakkary & Maestri (2008) reveal how children's emplacement within ad hoc acts of (re) appropriation and (re)use of household items in everyday domestic routines exposes them to creative ways of bringing material things together. Amongst my participants, everyone pointed to the home as the first key site of making-related learning. Participants felt strongly that they benefitted from exposure to making practices from a young age, even if they were not direct participants. This often related to seeing parents engage in a range of domestic tasks and hobbyist practices around the home, including sewing, knitting, embroidery, car maintenance, carpentry and household electrics. As a result they witnessed the value placed on both manual skill and the object(s) concerned, as well as having some insight into the processes, tools, and materials' capabilities. The home environment was, thus, for most participants highly conducive to a wide range of making opportunities, even if only as an observer in the first instance.

Discussion of sites outside the home reflected a fundamentally changing spatiality of making. Whilst on the one hand the rise of makerspaces and 'fab labs'[10] is argued by some to be a democratising force in widening public access to making opportunities (Anderson, 2012), other non-domestic making-places are argued to be disappearing. Barbara pointed specifically towards the loss of evening classes and the closing or merging of craft guilds.[11] Yet Barbara's argument needs to be considered alongside the emergence of new forms of craft-based sociality, such as 'knit and natter' (also known as 'stitch and bitch') groups for knitters (Minahan & Cox, 2007). Participants were clear that the formal education system was not the optimal location for engendering interest in making. Oscar (late 20s, software engineer), Nina and Barbara all felt that some exposure to making within the context of formal education was important, but there also needed to be informal opportunities from which people could select. The younger participants in particular (Oscar; Nina; Annie (early 20s, student); Mark (early 30s, organic farmer)) noted how the opportunities presented by formal education are not always embraced at the time.

Commenting on her experience of learning to sew in garment factories in Trinidad, Prentice (2008) suggests that the location in which making is learned and practised shapes the meaning and interpretation of that practice both in that moment of practice and subsequently. In other words, whether the making-place is a workplace, a leisure place, an educative place, or a place where other meanings dominate may influence how individuals interpret and internalise the activity performed therein – for instance, whether it is experienced as enjoyable, monotonous or relaxing. If wider societal engagement with making (in all its forms) is deemed both environmentally and economically important (as Carr & Gibson, 2016 compellingly argue), such considerations are worthwhile. To this end, Annie also highlighted the need for greater exposure of the increasing diversity of making practices within the wider socio-cultural context, particularly to encourage interest from younger audiences. This has perhaps started to emerge in recent years through growing television coverage and popular periodicals related to making and craft activities (Hall & Jayne, 2016).[12]

Digital places also played an important role for all participants, with most referring to Google and YouTube as key tools in their quests to learn new skills or problem-solve. Henry (early 30s, entrepreneur) stated that, when it comes to making, "There is no problem in the world to do with making that you can't solve with YouTube." Beyond David Gauntlett's excellent *Making is Connecting*, literature on the role of digital space as an information repository for craft and hobbyist makers has remained scant so far (although see Von Busch, 2010). As such there is considerable scope to explore how amateur makers juxtapose digital and physical spaces-of/for-making in their projects. Nina, Annie and Oscar agreed that YouTube can often provide enough information to get a project started, but questioned how far it could support a project's progress through to completion, including troubleshooting some of the more idiosyncratic challenges that may be encountered. Under these circumstances it was usually deemed necessary to turn to someone for help.

People

Beyond the important role played by friends and family in exposing the participants to making as a valued practice, it was clear that an explicitly social dimension underpinned much making activity. Several respondents (Simon, Hilary, Nina, Mark, Oscar, Norman) signposted friends, family and acquaintances as their key ports of call for help with projects they did not have all the skills or knowledge for, with others – such as Bob – being the person to whom others turned. Noting the extent to which he has learned crucial making and repairing skills from others in the farming community, Mark underlined the importance of building and maintaining relationships within one's specific community of maker-menders. In doing so, he emphasised the importance of access to expertise that is situated within a shared land-scape-of-making, whether that landscape comprises fields and hillsides (e.g. Krzywoszynska, 2015), motorcycle repair shops (e.g. Crawford, 2009), or homes (digitally connected or otherwise, e.g. Luckman, 2013). Whilst in part this is simply about identifying who can be called upon for advice and assistance, it also helps to build cohesion and trust within that community.

There is a valuable point to be drawn from this emphasis on the sociality of making related to the role these relationships can play in building resilience – both personal, in terms of self-efficacy ("I can do it!"), and at a community level, in terms of shared problem-solving ("We can do it!"). Further, this collaborative dimension of making indicates how maker-habitus is built in part through learning directly from (or with) others (Krzywoszynska, 2015 illustrates this neatly through her consideration of viticulture). Being able to develop making skills face-to-face and in conversation maximises the ability to transfer tacit knowledge through gesture, bodily sensation, corrective direction ("not like that, like this"; see for example O'Connor, 2007), when the 'tacit-ness' of the action means important nuances may get lost commu-nicating via other means, even a detailed instructional YouTube video.

Even for those participants for whom the social dimension of making was not a primary motivator, they saw – and sought to capitalise on – its benefits. Norman (early 50s, software engineer) primarily joined the makerspace to provide his young son with somewhere to develop his passion for electronics. Although Norman did not view himself as a 'maker' (and only really used the makerspace for hot-desking), he had decided to use his access to the tools and expertise to recondition an old record turntable. He stated his intention to send an email around the makerspace membership asking for help. Thus despite not intending to use the space as a maker himself, Norman's knowledge of the potentialities of the makerspace (its equipment and its knowledgeable members) drew out of him an inclination to (re)make (or repair/recondition) which may otherwise have remained dormant. His actions highlight one of the important potentialities of social spaces of making – the potential for 'non-makers' to be inadvertently drawn into a practice which may present them with unanticipated benefits (cf. Hitchings, Collins & Day, 2015).

For others the desire for social interaction preceded the making. This approach was more common amongst the female participants.[13] Nina began spending social time with friends attending craft classes and learning how to make dresses that fitted their bodies better than shop-bought garments (see also Hall & Jayne, 2016). Drawing on her years of experience of both formal and informal settings for learning making skills, Barbara argued that, "People don't want certificates, they want the social aspect." Her claim supports extant research into histories of craft-based socialities, which emphasise both the importance of such gatherings for (often gender-based) solidarity, and the utility of such gatherings for the sharing of skill, infor-mation and materials (Gauntlett, 2011; Hall & Jayne, 2016; Shercliff & Twigger Holroyd, 2016).

It was clear that recognition and esteem were also key motivators. Nina and her friends put photos of the dresses they had made on social media, which inspired enquiries and enthusiasm from other friends. Bob stated that, for him, recognition within a community of 'makers' was important to him, although in his case this was more about feeling a sense of belonging amongst "people of the same kind". Sarah Davies, in her research into leisure time spent in US makerspaces, identified that users see themselves as being of the same 'spirit' (Davies, 2017). This sense of solidarity, and associated sense of trust, helps to explain the value that has been attributed to shared spaces of making for individual wellbeing. This is exemplified, for instance by the Men's Sheds movement, which provides workspaces for isolated older men, helping to tackle loneliness as well as other mental and physical health issues (see, for example, Wilson & Cordier, 2013). Beyond the wellbeing benefits, these experiences hint at a (re)ignition of interest, not just in productive leisure time, but in 'socio-productive' leisure time linked (often through social media) to the psycho-social need for recognition within (and beyond) one's community. When contrasted with the lack of fulfilment that purportedly results from increasingly dematerialised social interactions, this raises important questions about how a drive for more meaningful social connection might be facilitated in ways that produce multiple sustainabilities.[14]

Thus far, the themes of play, place and people have been revealed as fundamental to the participants' orientations to the material world and, more specifically, to their readiness to manipulate some of its component parts through diverse practices of making. A licence to 'play' permits the development of a varied bank of embodied knowledge. Learning within a material place – primarily the home, but also a work environment or community space – seems to have a particularly profound influence on the ignition of interest and desire to build skill. However, people (family members, friends and acquaintances) are needed to make those places, and the practices which occur within them, meaningful and thus desirable for new practitioners to take them up. Together, these constitute the socio-material environment within which an enjoyment of – and thus disposition towards – making is engendered, and the tacit knowledges fundamental to maker-competence are developed. What, then, does this mean for how these 15 individuals responded to the consumer-cultural world? Their comments suggest two key realms of action in which a maker-habitus may fruitfully challenge unsustainable consumption. The first considers the tackling of built-in obsolescence by taking items apart to understand how they work, why they fail, and how functionality might be restored. This is discussed in terms of 'opening the black box' of consumer goods. The second focuses on how making skills need not – and, as the participants demonstrate, do not – exist in hobbyist siloes; rather, they can be put to work in a range of everyday contexts. This is discussed in terms of the transposability of knowledge.

Opening the 'black box'

The motto of *MAKE* magazine is, "If you can't open it, you don't own it" (Jackson, 2010: p.9). Participants' experiences of making (and resultant sense of competence) played a fundamental role in their ability to open up the 'black box'[15] of consumer goods. Common linkages were made between 'black-boxing' and blasé consumer attitudes towards object replacement rather than maintenance or repair. As Jim (late 50s, computer engineer) noted, obscuring the workings of material things impedes understanding of how they work. As a result, consumers may be intimidated by the idea of repairing them, or even taking them apart, thus inhibiting curiosity and preventing learning. Other participants reported how, as children, they had been actively encouraged to disassemble their possessions in order to facilitate this learning. Ken (early 60s, semi-retired engineer), reflecting on his childhood experiences, said, "If I wanted to saw in half my brand new toy in order to, to make something, nobody said 'you can't do that'..." Henry, similarly, was actively encouraged by his father to use all the power tools they had at home, and to take apart a broken record player bought from a flea market.

Hilary (early 50s, careers counsellor) and Annie both noted the empowerment that results from taking things apart, either as a means of attempting a repair, or in order to "cannibalise" unusable items for parts (as Hilary did with jewellery that she then remade into a 'new' item). Annie noted that, when faced with an unusable item:

> You've got a broken thing anyway, what are you going to lose by having a go at it? You either do it fine and it looks alright, you do it great and it looks brilliant, or you fluff it up and you have to give it to someone else to fix. You've at least had a go and you might have improved your skill by having a go.

Henry suggested that most people were "scared" of what was inside their consumer goods because of the fact that it was not only purposely hidden from them, but because this enclosure was emphasised by the threat of voiding warranties if attempts were made to gain access. He proposed his solution:

> I think every person should open one thing, and it doesn't matter what they open [...] take your old Walkman from the 1980s... take it apart... and that should be enough, just by the fact that you get a screwdriver and take it apart you can see, "oh there's the motor, I can see what that does, there's the battery...".

However, he also identified one key barrier: "The fact that I need a special screwdriver to open my iron, that prevents 95% of people from opening things up."

By opening up the 'black box', these participants opened themselves up to the potential of materials and their affordances, and, in doing so, added to their repository of embodied knowledge. This could then be called into action when attempting to diagnose and fix problems, whether a faulty kettle switch, a jammed necklace clasp, or, in Oscar's case, stereo speakers which emitted smoke. His existing embodied knowledge allowed him to diagnose the problem. When a second problem occurred, he was unable to fix it but he "learned a bit more about the components by at least having a good look."

Transposable knowledge

The second realm of action concerns the transference of embodied knowledge and skill from one context to another. Oscar had never attempted to transfer his knowledge of electronics onto speakers before, but he was willing to try, confident in his existing competence and the fact that he would add to his embodied knowledge whatever the outcome. Henry, frustrated that he could not find the correct tool to attempt to repair his iron, said, "If they [makerspace] don't have the screwdriver maybe I should just make the screwdriver." This ability to see potential "solutions" comes, according to Barbara, "only with experience", and is a clear example of how the embodied, tacit knowledge of the maker-habitus results in an ability to perceive the affordances in material things.

The most significant barrier to the acquisition of this experience is time. Problem-solving invariably takes longer in a 'new' context, particularly if some of the materials are unfamiliar. All participants in this project saw that they could transfer their making skills into contexts of practical everyday repair around the home (for example), but the overriding question was whether they considered doing so to be 'worth it'. Mike, Simon and Henry all stated that they possessed enough embodied knowledge to identify quickly whether or not they would be able to conduct the repair easily, quickly, and cheaply. If they sensed the repair would be neither easy, quick, nor cheap, they would be unlikely to attempt it despite possessing all the requisite knowledge, opting to buy a replacement instead.

One fundamental tension inherent to the notion of transferring skill between domains connects with the changing spatiality of making. Melissa (early 40s, librarian) described her concern that the making activity that took place at the makerspace she belonged to would only ever exist in that silo; that it would remain separate from 'real life' – her family's home life. Having two young children who each favoured very different forms of creative expression, she was particularly keen to avoid framing making as an activity only possible in a particular place at a particular time. Whilst this view was driven by her concerns for her sons' creativity, she also mused on the extent to which the wider (adult) makerspace membership might similarly contain their making activity within that designated space. She argued that setting making apart as a distinct practice risks producing a sub-culture that the majority feel is not relevant to them. Making therefore becomes a privileged, exclusionary activity. In light of the age and gender implications of the changing spatiality of mending, and in spite of the claims about the inherent 'democracy' of the maker movement (Anderson, 2012), there are questions to be answered about the inclusivity and accessibility of making practices and places.

In demonstrating their willingness – even eagerness – to open up the 'black box' of everyday consumer goods, and in exemplifying the transposability of maker-competence from one domain to another, these 15 makers have highlighted the role played by their making experiences in challenging some of the strictures imposed by a consumer culture predicated on rapid obsolescence. In the conclusion to this chapter, I reflect on what this suggests about the potential of the maker-habitus as a theoretical lens and as a driver of more widespread support for making (and more sustainable everyday consumption) within the wider socio-cultural sphere.

Conclusion: a sustainable future in the making?

> ... the more you do this, the more you know, the less you want to waste...
>
> (Mark)

Recent years have seen resurgent interest in a wide range of making practices. New tools (such as laser cutters) and forms of connectivity (the internet) have shaped the image, visibility and spatiality of those practices, with claims about the democratisation and political potential of making (though sometimes contestable) being the result. In this chapter I have linked this purportedly democratic and political potential of the so-called 'maker movement' to the ongoing quest for more sustainable everyday consumption. Growing public interest in acquiring making skills presents opportunities to throw back the opaque veil of the fetish imposed by global consumer culture, and instead reconnect makers (novice, amateur, experienced and professional) with the materials and labour of production, and, in turn, with the environment from which all material things come and to which they must, in some form, return.

In support of this argument I conceptualised the maker-habitus as a means of explaining how making practice might, through the genesis of a particularly materially sensitive disposition, enable the perception of un(der)used, un (der)loved, or just plain broken, material things as 'vibrant matter' (Bennett, 2010) – things with the potential for reuse, thus avoiding the waste stream and negating the need for new acquisitions. At the heart of this concept of maker-habitus is the notion that, by accumulating tacit, embodied knowledge through cumulative experiences of making, makers might be particularly well-placed to perceive the affordances of material things. The ethnographic work drawn upon here has evidenced not only how the maker-habitus forms (literally) in practice (in play, in place, and with people), but its salience has been illustrated by the participants' confidence in applying their knowledge to the black boxes, obsolete objects and functional failures that characterise growing numbers of consumer cultural items.

Nevertheless, the claims made here for the virtues of making as an important ally in the quest for more sustainable consumption are not without their tensions. Is making really the friend of sustainability when, as freely admitted by Simon, Mike and Bob, part of the joy – the playfulness – of making comes from making "pointless machines"? Whilst the inherent tensions related to the use of resource (materials, components, energy) cannot be overlooked, at the same time – and as these men demonstrated – their experiences of making "pointless machines" resulted in a maker-habitus which was clearly orientated towards repairing, maintaining, avoiding waste and, more broadly, finding creative solutions. In that sense, it might be argued that the resource 'costs' are compensated (or at least balanced) by resources saved in the course of repairs achieved, as well as repairs (or similar) achieved by others inspired or mentored by these competent, confident makers.

A second tension emerges in response to the claims for democratic access associated with contemporary making culture. Whilst the spread of digital technologies has, undoubtedly, given some people greater access to making – whether to learn skills or hone them – others seem to be losing out. The emergence of fab labs and makerspaces has created a valuable new set of spaces for types of making that require access to expensive technical (and often very large) equipment. Not all charge membership or joining fees; some charge on a 'per use' basis for the equipment. Nevertheless, cost barriers will constitute a barrier to some. A second key barrier, which both relates to makerspaces and fab labs and to the changes in contemporary making more broadly, concerns intersections of age and gender. Makerspaces and fab labs remain significantly skewed towards male membership, in large part because the kinds of resources on offer tend to speak to stereotypically 'male' interests.[16] In stark contrast, and as noted by Barbara, many of the making networks most accessible (and appealing) to older women, such as craft guilds and evening classes, are being lost. Whilst this may simply reflect naturally occurring demographic shifts, it should highlight the need for awareness in ensuring all kinds of making constituents have a place where their interests can be nurtured.

Molly Scott-Cato (2014) argues that there is much that sustainability education can learn from craft education. Specifically she points to learning-by-doing, an acceptance – even embrace – of messiness and error, and the importance of learning within a community in order to embed the transformation in action that is initiated. This corresponds closely with the arguments made here. As Annie noted:

> For your typical person [...] there's going to be some apathy when it comes to sustainability, but if you say here's a project you can do, you can involve your kids [...] oh, and it also does this [ticks sustainability boxes], then it's interesting, it's useful.

In short, making may provide another opportunity for the 'sustainability by stealth' or 'inadvertent environmentalism' (Hitchings, Collins & Day, 2015) that seems to have become more achievable since the recession of 2008 pushed sustainability down household agendas. As Melissa stated, "You have to have people do it – give them the bug." Harnessing making to sustainability may mean the public gets two bugs... and manifold benefits.

Notes

1 This is due to the socio-economic inequalities that prevent equal access to opportunities to learning, tools, information and mentorship – all of which, as this chapter argues, are fundamental to participation in, enjoyment of and competence in making practices.
2 Although some would question whether these practices ever 'went away' (e.g. Hackney, 2013), their appeal to an increasingly diverse demographic suggests that a social, cultural and political shift worthy of acknowledgement is underway.
3 Gelber (1999) argues that the counterculture of the 1960s saw the emergence of a new arts and crafts movement similarly underpinned by an anti-capitalist ideology.
4 Gibson's work develops ideas of 'Umwelt' (subjective perception of 'the-self-in-the-world') first discussed earlier in the twentieth century by biologist and biosemiotician Jakob von Uexküll.
5 I write 'E/environmental' with both upper and lower case here to convey that these implications may affect the 'small e' environment of an actor's immediate surroundings and/or the 'big E' Environment which underpins concerns with sustainability.
6 A significant value-action gap (see, for example, Barr, 2006) exists here, between expressions of anxiety about the waste associated with, for example, fast fashion (e.g. Morgan & Birtwistle, 2009), and the relative inaction amongst consumers to tackle it.
7 All participants have been attributed pseudonyms.
8 From 10.00–12.00 on Sunday mornings makerspace members could bring their children to learn making skills in the context of a project. Members take it in turns to facilitate the session and share their specific expertise.
9 'Modding' is an abbreviation of 'modifying', and refers to remaking practices where material objects take on new material forms, and often new purposes, in order to improve or extend their usability.
10 A 'fab lab' (abbreviation of 'fabrication laboratory') is broadly synonymous with a makerspace.
11 There are clear age and gender dimensions – and thus implications – to these changes, which, whilst outside the scope of the present discussion, warrant exploration.
12 TV programmes include the BBC's *Great British Sewing Bee* and *Great Pottery Throw Down*, Channel 4's *Kirstie's Homemade Home*, as well as Freeview channel Quest's series *How It's Made*. Popular periodicals include *MAKE* magazine, *Love Sewing*, *Crafts Beautiful* and *Vogue Knitting*.
13 Regrettably a more detailed consideration of the gendered dimensions of sociality and making is outside the scope of the present discussion. Its salience as a research focus is indicated by, amongst others, Luckman (2013); Hall & Jayne (2016); Jackson (2010); Wilson & Cordier (2013).
14 For instance, community courses that provide social connections, practical skills, and a sense of the environmental embeddedness of those social relationships and material actions.
15 This term refers to the now common practice of physically preventing access to the workings of a wide range of consumer goods, from cars to mobile phones.

Purportedly driven by concerns for consumer safety, one of the results is growing ignorance amongst consumers about how things are made and how they might be maintained or repaired.

16 In my 18-month ethnographic work with the makerspace discussed here, I was the only female in the building around 40% of the time.

References

Anderson, C. (2012) *Makers: The New Industrial Revolution.* New York: Crown Business.

Barr, S. (2006) Environmental action in the home: investigating the 'value-action' gap. *Geography,* 91(1), 43–54.

Bauder, H. (2005) Habitus, rules of the labour market and employment strategies of immigrants in Vancouver, Canada. *Social and Cultural Geography,* 6(1), 81–97.

Bennett, J. (2010) *Vibrant Matter: A Political Ecology of Things.* Durham, NC: Duke University Press.

Bourdieu, P. (1984) *Distinction: A Social Critique of the Judgement of Taste.* London: Routledge.

Brook, I. (2012) Make, do, mend: solving placelessness through embodied environmental engagement. In Brady, E. and Phemister, P. (eds) *Human-Environment Relations: Transformative Values in Theory and Practice.* Dordrecht/Heidelberg/London/New York: Springer Netherlands, pp. 109–120.

Campbell, C. (2005) The craft consumer: culture, craft and consumption in a postmodern society. *Journal of Consumer Culture,* 5(1), 23–42.

Carr, C. and Gibson, C. (2016) Geographies of making: rethinking materials and skills for volatile futures. *Progress in Human Geography,* 40(3), 297–315.

Castells, M. (2012) *Aftermath: The Cultures of the Economic Crisis.* Oxford: Oxford University Press.

Comor, E. (2010) Digital prosumption and alienation. *Ephemera,* 10(3/4), 439–454.

Cooper, T. (2005) Slower consumption: reflections on product life spans and the 'throwaway society'. *Journal of Industrial Ecology,* 9(1–2), 51–67.

Craggs, R., Geoghegan, H. and Neate, H. (2013) Architectural enthusiasm: visiting buildings with the twentieth century society. *Environment and Planning D,* 31(5), 879–896.

Crawford, M. (2009) *The Case For Working With Your Hands.* London: Penguin.

Cushion, C. J. and Jones, R.L. (2014) A Bourdieusian analysis of cultural reproduction: socialisation and the 'hidden curriculum' in professional football. *Sport, Education and Society,* 19(3), 276–298.

Dant, T. (2005) *Materiality and Society.* Milton Keynes: Open University Press.

Dant, T. (2010) The work of repair: gesture, emotion and sensual knowledge. *Sociological Research Online,* 15(3).

Davies, S. (2017) Characterizing hacking: mundane engagement in US hacker and makerspaces. *Science, Technology and Human Values.* DOI: https://doi.org/10.1177/0162243917703464

Eden, S. and Bear, C. (2011) Reading the river through 'watercraft': environmental engagement through knowledge and practice in freshwater angling. *Cultural Geographies,* 18(3), 297–314.

Fox, S. (2014) Third wave do-it-yourself (DIY): potential for prosumption, innovation, and entrepreneurship by local populations in regions without industrial manufacturing infrastructure. *Technology in Society,* 39, 18–30.

Frow, J. (2003) Invidious distinction: waste, difference and classy stuff. In Hawkins, G. and Muecke, S. (eds) *Culture and Waste: The Creation and Destruction of Value.* Oxford: Rowman and Littlefield, pp. 23–38.

Gamble, J. (2001) Modelling the invisible: the pedagogy of craft apprenticeship. *Studies in Continuing Education,* 23(2), 185–200.

Gauntlett, D. (2011) *Making Is Connecting.* Cambridge: Polity.

Gelber, S.M. (1999) *Hobbies: Leisure and the Culture of Work in America.* New York: Columbia University Press.

Geoghegan, H. (2013) Emotional geographies of enthusiasm: belonging to the Telecommunications Heritage Group. *Area,* 45(1), 40–46.

Gibson, J.J. (1986) *The Ecological Approach to Visual Perception.* Hillsdale, NJ: Lawrence Erlbaum Associates, Inc.

Graham, S. and Thrift, N. (2007) Out of order: understanding repair and maintenance. *Theory, Culture and Society,* 24(3), 1–25.

Gregson, N., Metcalfe, A. and Crewe, L. (2009) Practices of object maintenance and repair: how consumers attend to objects within the home. *Journal of Consumer Culture,* 9(2), 248–272.

Hackney, F. (2013) Quiet activism and the new amateur: the power of home and hobby crafts. *Design and Culture,* 5(2), 169–193.

Hall, S.M. and Jayne, M. (2016) Make, mend and befriend: geographies of austerity, crafting and friendship in contemporary cultures of dressmaking in the UK. *Gender, Place and Culture,* 23(2), 216–234.

Hatch, M. (2013) *The Maker Movement Manifesto: Rules for Innovation in the New World of Crafters, Hackers, and Tinkerers.* Columbus, OH: McGraw Hill Professional.

Hitchings, R., Collins, R. and Day, R. (2015) Inadvertent environmentalism and the action-value opportunity: reflections from studies at both ends of the generational spectrum. *Local Environment: The International Journal of Justice and Sustainability,* 20(3), 369–385.

Ingold, T. (1992) Culture and the perception of the environment. In Croll, E. and Parkin, D. (eds) *Bush Base, Forest Farm: Culture, Environment and Development.* London: Routledge, pp. 39–55.

Ingold, T. (2000) *The Perception of the Environment: Essays on Livelihood, Dwelling and Skill.* London: Routledge.

Ingold, T. (2013) *Making: Anthropology, Archaeology, Art and Architecture.* London: Routledge.

Jackson, A. (2010) Constructing at home: understanding the experience of the amateur maker. *Design and Culture,* 2(1), 5–26.

Kopytoff, I. (1986) The cultural biography of things: commoditization as process. In Appadurai, A. (ed.) *The Social Life of Things: Commodities in Cultural Perspective.* Cambridge: Cambridge University Press, pp. 64–91.

Krzywoszynska, A. (2015) What farmers know: experiential knowledge and care in vine growing. *Sociologia Ruralis,* 52(2), 289–310.

Lindtner, S. (2015) Hacking with Chinese characteristics: the promises of the maker movement against China's manufacturing culture. *Science, Technology and Human Values,* 40(5), 854–879.

Luckman, S. (2013) The aura of the analogue in a digital age: women's crafts, creative markets and home-based labour after Etsy. *Cultural Studies Review,* 19(1), 249–270.

Maller, C., Horne, R. and Dalton, T. (2012) Green renovations: intersections of daily routines, housing aspirations and narratives of environmental sustainability. *Housing, Theory and Society*, 29(3), 255–275.

Mann, J. (2015) Towards a politics of whimsy: yarn bombing the city. *Area*, 47(1), 65–72.

McDonough, P. (2006) Habitus and the practice of public service. *Work, Employment and Society*, 20(4), 629–647.

Merleau-Ponty, M. (2014 [1981]) *The Phenomenology of Perception*. London: Routledge and K. Paul; New York: Humanities Press.

Minahan, S. and Cox, J.W. (2007) Stitch 'n bitch: cyberfeminism, a third place and the new materiality. *Journal of Material Culture*, 12(1), 5–21.

Mitchell, S. (2011) *Objects in Flux: The Consumer Modification of Mass-produced Goods*. Unpublished PhD thesis. School of Architecture and Design, RMIT University.

Morgan, L.R. and Birtwistle, G. (2009) An investigation of young fashion consumers' disposal habits. *International Journal of Consumer Studies*, 33, 190–198.

O'Connor, E. (2007) Embodied knowledge in glassblowing: the experience of meaning and the struggle towards proficiency. *The Sociological Review*, 55 (s1), 126–141.

Patchett, M. (2016) The taxidermist's apprentice: stitching together the past and present of a craft practice. *Cultural Geographies*, 23(3), 401–419.

Paton, D.A. (2013) The quarry as sculpture: the place of making. *Environment and Planning A*, 45, 1070–1086.

Pimlott-Wilson, H. (2011) The role of familial habitus in shaping children's views of their future employment. *Children's Geographies*, 9(1), 111–118.

Polanyi, M. (1958) *Personal Knowledge: Towards a Post-critical Philosophy*. London: Routledge and Kegan Paul.

Prentice, R. (2008) Knowledge, skill, and the inculcation of the anthropologist: reflections on learning to sew in the field. *Anthropology of Work Review*, XXIX (3), 54–61.

Price, L. (2015) Knitting and the city. *Geography Compass*, 9(2), 81–95.

Ritzer, G. (2014) Prosumption: evolution, revolution, or eternal return of the same? *Journal of Consumer Culture*, 14(1), 3–24.

Scott-Cato, M. (2014) What the willow teaches: sustainability learning as craft. *Learning and Teaching*, 7(2), 4–27.

Sennett, R. (2009) *The Craftsman*. New Haven, CT: Yale University Press.

Shercliff, E. and Twigger Holroyd, A. (2016) Making with others: working with textile craft groups as a means of research. *Studies in Material Thinking*, 14, paper 07.

Smyth, E. and Banks, J. (2012) 'There was never really any question of anything else': young people's agency, institutional habitus and the transition to higher education. *British Journal of Sociology of Education*, 33(2), 263–281.

Straughan, E. (2015) Entangled corporeality: taxidermy practice and the vibrancy of dead matter. *GeoHumanities*, 1(2), 363–377.

Toffler, A. (1980) *The Third Wave*. New York: William Morrow and Company.

Torrone, P. (2005) Mod your pod: enhance your iPod with a Linux upgrade. *Make: Technology On Your Time*, 2, 135–137.

Von Busch, O. (2010) Exploring net political craft: From collective to connective. *Craft Research*, 1, 113–124.

Wacquant, L. (2005) Habitus. In Beckert, J. and Zafirovski, M. (eds) *International Encyclopedia of Economic Sociology*. London: Routledge.

Wakkary, R. and Maestri, L. (2008) Aspects of everyday design: resourcefulness, adaptation, and emergence. *International Journal of Human-Computer Interaction*, 24(5), 478–491.

Wilson, N.J. and Cordier, R. (2013) A narrative review of Men's Sheds literature: reducing social isolation and promoting men's health and well-being. *Health & Social Care in the Community*, 21(5), 451–463.

Yarwood, R. and Shaw, J. (2010) 'N-gauging' geographies: craft consumption, indoor leisure and model railways. *Area*, 42(4), 425–433.

12 Everyday *Kintsukuroi*

Mending as making

Caitlin DeSilvey and James R. Ryan,
with photographs by Stephen Bond

To think about repair requires us to recognize our own failures and imperfections and those of the world we live in, to take seriously what we may be unreflectively inclined to regard as the necessary but uninventive and uninspiring work of repairing the damage due to such flaws. It means attending to properties in things – their repairability – and capacities in individuals – their talents for mending – towards the atrophy of which there appear to be powerful economic incentives.

(Spelman, 2002: p.138)

Introduction

Kintsukuroi (or *kintsugi*) is the Japanese art of repairing cracks in broken pottery with gold, silver or platinum lacquer. It also expresses the idea that breaking and mending can be an important part of the life of an object, adding to its beauty and meaning. Although this concept has its origins in a cultural context far from the back street repair workshops that we focus on in this chapter, there is, as we hope to show, a clear resonance with the work carried out in these places. In 2010 we began a collaborative research project that brought together two cultural geographers and a photographer, Steve Bond, to document the visual and material cultures associated with the making and mending of everyday objects in southwest England. We named our venture 'Small is Beautiful?' in gentle deference to E. F. Schumacher's classic 1973 collection of essays, a text that championed the urgent need for human societies to forge forms of living that were more economically, socially and ecologically sustainable. Our project aimed not only to record the material cultures associated with the practice of mending ordinary objects, but also to test and refine collaborative methods for the integrated investigation of visual, material, and social relationships. In addition, we wanted to engage academic and non-academic communities in conversations about everyday aesthetics, cultural value, and economic sustainability.

In this chapter we chart the project as it evolved, reflecting on the social and political moment in which the research was placed, and sharing some of the insights that arose from reception of the images that we created. We also

explain our commitment to the photograph as something *made* through the use of specific tools and expertise, and draw out the parallels between our making and the forms of making we encountered in the places we documented. Finally, we consider how collaborative practice – in this instance geographers working with a photographer – can illuminate rich and embodied fields of action in which the boundaries between material objects and those who make, repair and appreciate them are continuously remade.

Geographies of mending

In both academic and applied contexts, practices of repair and maintenance have often been eclipsed by a focus on making and innovation, a bias evident in fields as diverse as tech culture (Chachra, 2015) and public infrastructure (Russell and Vinsel, 2017). Recently, however, academics and activists have begun to establish repair and mending as a vibrant field of enquiry (Graham and Thrift, 2007; Jackson, 2014; Lepawsky et al., 2017), and also to expose the ways in which repair must be understood as a creative practice in its own right (Bond et al., 2013). Some of this recent work focuses on less developed economies in the global South (as well as poorer communities in affluent societies), where the salvage, repair and creative reuse of material objects remains an essential survival strategy rather than a lifestyle choice (Callén and Criado, 2015; Houston, 2017). In contrast, practices of mending have been relegated to the periphery of productive economic activity in developed economies, forced into the margins by the sheer scale and dominance of mass consumer culture. Consumers who once regarded purchased objects – from clothes to computers – as worthy of maintenance and repair, now widely accept them as entirely disposable (Cooper 2005; Van Nes 2010). Moreover, this disposability relies on a wider division of labour based on flows of end-of-life goods from the global North to the global South (Crang et al., 2013).

Partly in response to the dominant culture of obsolescence and disposability, over the last decade a range of social movements have championed the revival of making and mending, with the formation of grassroots initiatives focused on the repair, reuse and 'upcycling' of material items hitherto regarded as waste (Janigo et al., 2017). Some of these movements – most often those associated with metropolitan, relatively privileged socio-economic enclaves – invest such activities with a political critique of capitalist society (Fickey, 2011; Bramston and Maycroft, 2014). The revival of interest in mending and repair is often (either consciously or not) linked to older seams of thought and practice: the 'head, hand, heart' sentiments behind the Arts and Crafts movement of the nineteenth century resurface on lifestyle blogs; the WWII 'make do and mend' mantra is fetishised by a new urban elite. Popular interest in repair is recursive, remerging at key points, such as in the counterculture of the 1960s, when Stewart Brand's *Whole Earth Catalog* (1968) elevated the virtues of 'hacking' over mundane and mainstream 'planning' (Morozov, 2014).

We began our research a couple of years after the 2007–2008 financial crisis, just as the hard realities of 'austerity Britain' and economic recession began to force many households to consider a return to 'make do and mend' out of necessity. As the project went on, we realised that our research coincided with a broader rekindling of interest in craft and making (Hackney, 2013; Thomas et al., 2013), and a rise in demand for the bespoke and handmade (Luckman, 2015). We also became aware of the emergence of new kinds of 'craftivism' and DIY hacker cultures, mediated by digital technologies and drawing in other communities of interest and expertise (Orton-Johnson, 2014). Although our research project was forged in the context of this upwelling of interest in making and mending, its focus was on low-profile, small-scale repair businesses, whose proprietors were generally unware of the repair revival taking place in wider popular culture.

Southwest England has few major urban centres, but an extensive network of craft-based and creative industries, making it a distinctive location in which to document cultures of repair (Luckman, 2012; Thomas, Harvey and Hawkins, 2013). Over two years, we made recurring visits to 20 small businesses in the region. We selected the businesses based on the type of work they carried out and their willingness to participate in the project. At a minimum, participation involved one visit from a photographer/geographer pair, and an

informal interview. We also invited business owners and employees to attend one of four 'public conversations', which were held in connection with exhibitions of Steve Bond's photographs. Altogether, representatives from 13 of the 20 businesses participated in at least one of these events, some speaking with researchers in front of an audience of local residents, shop customers, artists, photographers and academics.

This project began with the simple aim to make local mending cultures visible. In doing so, as with the Japanese mending tradition of *kintsukuroi*, we wished to highlight, rather than disguise, the complex biographies of objects (Keulemans, 2016). In embarking on this project, we recognised the inadequacy of the overgeneralised concept of the 'throwaway society' (Gregson et al., 2007). Rather than following the linear trajectories of objects from consumption to disposal, we wanted to understand the range of different values associated with repair practices. Emotional value, relational value, aesthetic value and ecological value are all expressed, in various ways, in the desire to mend a broken object. In seeking repair, people are guided by concerns for thrift and durability, but they also place value on non-commoditised aspects of repair, including the workplaces where it is carried out and the social relations embedded in these places. In this chapter, we focus on two themes that emerged from the research – the parallels between our research practice and the practices of making and mending that we encountered in the field; and way in which a shared spirit of 'making' animated our public conversations and encouraged people to reflect on wider issues of cultural value, social cohesion and economic change.

Research-craft and the expanded exhibition

The 'Small is Beautiful?' project originated at a moment in cultural geography when discussions about the relationships between geography, visual culture and art were beginning to engage more fully with the potential of collaborative practice (Hawkins, 2014). It was becoming commonplace for geographers to work alongside artists and integrate creative visual methodologies into their work. This project sought to extend such work, drawing on visual and sensory methodologies pioneered in anthropology (Pink, 2009) and responding to calls in geography for 'visual and material research that unravels, disturbs and connects with processes, embodied practices and technologies' (Rose and Tolia-Kelly, 2012: p.3). In this project, we chose to adopt an approach that treated the photograph not as an art object – produced primarily for visual consumption and contemplation – but as a crafted object, made for a particular purpose, to be (as Steve was fond of reminding us) both 'useful and beautiful'. Like the people in the repair workshops we encountered, we made choices about materials and processes, and applied (and acquired) skills through the process of completing certain tasks. Steve's expertise in the photographic craft was essential to this process, and we wanted to make this visible. As it happened, this element of the work became central to the exhibition of the photographs, and the quality and content of the conversations they provoked.

When we introduced our project to potential participants, we initially explained that we were interested in documenting places where ordinary objects were repaired. We quickly realised, however, that the people who were

receptive to our project thought of what they did as much more than simple 'repair'. For these people, mending was a complex practice that integrated elements of problem-solving and invention, as well as, often, community service and social work. One of our menders told us: 'What's a repair? It could be a modification, or an improvement...I respond to problems...the problems of today...I look at a problem and consider all of the options in my repertoire.' Influenced by this perspective, we began to think about the repairers we were encountering as skilled practitioners of 'craft' (Adamson, 2007), keepers of tangible and tactile skills lost to many workers in modern knowledge economies (Sennett, 2009; Crawford, 2010). Steve's photography evolved in response to this awareness. As photographic maker, Steve documented an object, setting or process that was already imbued with creative potential by its owner and handler. By the end of the project, the repaired objects we had encountered included: shoes, clothes, books, sewing machines, motorcycle seats, ironwork, clocks and watches, typewriters, small electrical appliances, musical instruments, bicycles, small engines, ceramics, and cane chairs.

After Steve had produced the photographs, we were faced with the task of ordering and classifying them for viewing and display. Rather than judging the photographs on their artistic merit or on stylistic grounds, we worked together in what, to an outsider at least, might appear to be a much more intuitive and fluid set of criteria. At the heart of such selection lay a kind of 'capacious aesthetics', an accommodation of feeling for the material and visual qualities of the images (Highmore, 2016). We would lay prints of the photographs on the floor or on a table and reorganise them in different configurations, circling the images waiting for particular images and sets of images to catch our eyes. As we became attuned to the affective qualities of Steve's photographs we found that certain images seemed to want to be together, in pairs or triplets, or in series, drawing out certain patterns and relationships (we termed our family groupings of photographs 'SiBlings').

Our exhibition strategy extended the craft sensibility, and focused on treating the exhibited photographs as objects in their own right, with a material as well as a visual presence in the world. For our first series – seven sets of three – we mounted the prints on 3mm aluminium. This involved complex deliberations with a printer in Exeter, and the prints were sent to Yeovil for mounting, where some random but necessary cropping occurred. We then created a set of 21 tiny shelves, using aluminium architrave with a conventional application in shower installations. A later set of five images was printed on large sheets of canvas. Sorting out the technicalities of hanging these 'flags' involved three return visits to an Exeter ironmongery. At the exhibitions, we encouraged people to touch and handle the photographs that were mounted on aluminium, to know them with their hands as well as their eyes. The mounted photographs, as well as the larger prints and the flags, accumulated signs of their movement through the world – scratches and dings, nicks and smudges.

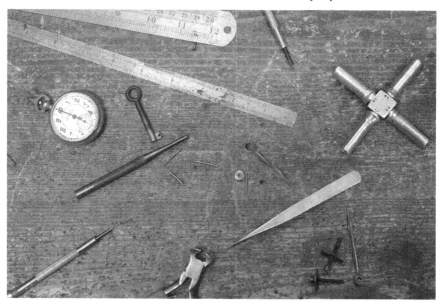

Taking our families of photographs to four exhibitions and two conferences over a year and a half – from Kendal, Cumbria to St Austell, Cornwall[1] – allowed us to share the images and witness the effect of their affective qualities on a wider audience of makers, menders, artists, academics and shop customers. Each exhibition prompted extended discussions about mending, menders and things. The project exhibitions were organized not merely to display the photographs, but as extended forms of visual ethnographic method. In this sense, the exhibitions functioned as creative events, part of the 'expanded creative register' (Hawkins, 2014: p.13) where artists, critics and audiences co-create sets of meanings around the photographs. The photographs sparked reflection on aesthetics and politics, nostalgia and future potential (Pink, 2013). These themes were held in tension at the moment of reception, as people stood before the images and made sense of them together, and talked about them in larger assembled groups.

In the exhibited photographs, repair shop owners saw their places of work in ways that they had never before done so, and found they could relate to them in new ways. They also were prompted to compare their own work and experience with that of other repairers. Close up photographs of tools, surfaces and objects elicited discussions about the properties of materials, and the appropriate tools for specific tasks. Menders often were moved to comment on the skill involved in surrendering themselves to the agency of the materials. It became clear that many menders see repair less as a straightforward process of imposing form on materials than as a series of interventions in what Tim Ingold terms 'fields of force and flows of material' (Ingold, 2009: p.91).

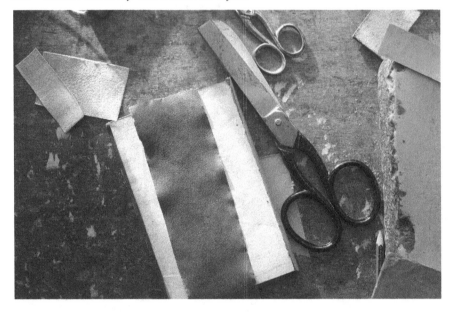

Making is not simply the human assertion of designed form onto passive material; rather, form is generated as a co-production in which human makers work with and are shaped by the animate, worked-upon matter (Ingold, 2013). The makers and repairers involved in this project often understood this process in quite intuitive and humorous ways. Some of the menders testified to how objects have an obdurate quality, one that might helpfully be captured in the concept of 'resistentialism,' a concept (coined originally by the British humourist and critic Paul Jennings) which has recently found new currency in renewed concern for the ways that inanimate objects resist humans' attempts to transform or work with them (Jennings, 1950; Elster, 2003). The craft of repair always involves, in some sense, the capacity to accommodate the independence of things (Hitchings, 2006).

Despite the absence of human bodies in many of Steve's photographs, the images capture the quality of the repair workshops as embodied spaces, and people at the exhibitions often commented on the sense of suspended animation expressed by the arrangement of tools and materials, as if the worker has just stepped away from the bench. These places have evolved around the needs of their human workers and the material qualities of the items they work with. Just as a quarry might be thought of as a sculpture in its own right (Paton, 2013), repair workshops are themselves sympathetic embodiments of repair, places whose surfaces, shapes, colours and smells reveal the accreted processes through which menders accumulate familiarity with materials, tools and objects. Steve's photographs responded to the distinctive quality of the tools and materials used in repair workshops, often highlighting

the sense of 'fluidity' that emerges between a tool, its operator and the space around it. Like Steve's camera, the repairers' tools are appropriate technologies, pressed into service of a range of requirements; in skilled hands they adapt in a fluid fashion to the demands placed upon them (De Laet and Mol, 2000). The photographs also exposed the capacity – often expressed by the participants in this project – for making and mending to provide a purposive engagement with the material world, in a tactile materialisation of both environmental and social values (Brook, 2012).

By giving close scrutiny to the sites and materials of repair, the photographs encourage us to recognise that repair is a process of creative intervention that does not seek to reverse entropic processes of decay, but merely postpones such inevitable processes, temporarily restoring an object's use, purpose and value to its owner (DeSilvey, 2006). Like the act of repair itself, the photographs command detailed attention to the material qualities of objects, while they also carry traces of cultural memory from the objects being repaired into a parallel visual and material register. The blacksmith who participated in the project told a story of a woman who came into his shop with an old copper jug that had been dented. The woman told him that her grandchildren had been visiting, and in a chaotic moment involving the dog, the cat, and children, the jug had been knocked off a shelf. She asked, 'Can you knock the dent out?' The blacksmith replied, 'Of course I can, but you will never tell that story again, because the dent is the trigger for the story.' The woman left the dent as it was. At the exhibitions and in the shops themselves, owners of shoes, clothes and bags proudly told us of how their much loved items had

been kept alive by the attentions of menders like those involved in the project, through countless patches and repairs. In this sense, repaired articles as well as the places of repair are 'assemblages' that emerge from networks of materials and entangled agencies (Edensor, 2011). Like photographs, acts of repair are gestures of temporary stabilisation, momentarily fixing material in flux and decay.

Some ethnographic research on consumer objects in the home has shown how practices of repair and maintenance are central to the processes whereby consumer objects assume their identities (Gregsonet al., 2009). Different kinds of restorative acts, from cleaning to full-scale repair, have varying effects on the status of consumer objects, and failure to provide maintenance and repair can result in an object's devaluation and disposal. Although this research project was not concerned with the position of consumer objects in the home, it showed that, for those who frequent repair shops, the repair of objects is often undertaken out of appreciation for their emotional and sentimental value – rather than a concern for their status as consumer objects. Often, these values are entangled with memory, and with the connection the object opens to past experience – a quality also attributed to the photographs. An audience member at the Bridport exhibition said, 'I can smell my grandfather's workshop when I look at these photographs.' Other people commented about the way the images triggered memories of tactile sensations, of other places and pasts. They wanted to share these memories, and the exhibitions became spaces for quite intimate exchanges, often between strangers. There were many animated conversations in front of the images on display, with people talking about the objects in the images, how they were made and used, and then moving on to discussions about other things – lamenting the 'throwaway society' and the decline of local shops, talking about their personal experience of cultural and economic change. Engagement with the photographs highlighted how the affective quality of repair workshops does not come simply from objects themselves but from the entire affective apparatus of the place and its contents, including its smells, atmosphere, colours, and the unpredictable arrangement of evocative 'stuff' (Anderson, 2014; Boscagli, 2014).

Making value, making relationships

Some anthropological scholarship posits a general distinction between 'Western' and 'non-Western' practices of exchange and value, and suggests that Western practices are generally characterised by impersonal relationships, where things are understood as inanimate commodities in monetised systems of exchange (Kopytoff, 1986). 'Non-Western' exchange practices, by contrast, are shown to exhibit revealing and often very personal entanglements between and among 'things' and 'people' (Napier, 2014). Yet, our investigations with people in southwest England who engage with repair showed that in this context the boundaries between 'things' and 'people' are cut through with emotional, affective and sensory connections. Several of the workshops we visited also functioned as rescue homes for temporarily abandoned objects. Stick of Lostwithiel housed a museum of rare shoes and other related items. Similarly, Sew-Quick in Falmouth (now moved to a new premises in Penryn), rescued various items from the brink of extinction (including an industrial iron and a 1970s sewing machine) – not in pursuit of financial benefit but for the love of these items and the pleasure taken in exercising restorative skill in their care. The transactions undertaken within repair workshops often have powerful socially integrative effects by fostering shared human appreciation and care for the material qualities and meanings of things. One of the menders involved in the project commented that he loved rescuing and reinventing objects that would otherwise have been discarded. He remarked, 'I love that side of it where you take something that's been beaten up and used and then turn it back into something that is usable again.' He also commented on his appreciation for the research, and its recognition of his work: 'I just think it's marvelous that you guys have…caught our vision for it…because quite often we're in little back street shops where people don't find us unless by chance.'

Over the course of the project we came to appreciate that people employed in mending and repair trades understand their relationship to the objects and materials they work with, and to the communities where they are located, primarily as one of service and vocation. They think of themselves as makers, inventors and creators, who specialise in the skilful manipulation of materials and take pleasure in keeping things alive; they actively resist characterisations of their work as being 'just repair' or as purely about financial exchange. Indeed, repair workshops often involve social transactions that confound bald

economic logic, certainly of the type generally found on the high street. Several loyal repair shop customers told us about how they often struggled to convince repairers to accept adequate compensation for their labour. One man told us how he had his vacuum cleaner totally rewired, but the shop proprietor 'only wanted £5 and wouldn't take any more'. When we first visited Stick of Lostwithiel, Caitlin's rucksack zip had just broken. The proprietor repaired it on the spot but refused to charge any more than the cost of the zip, prompting Caitlin to purchase a £4.60 container of NikWax. It quickly became clear to us that the activities in repair shops frequently involved values that were not recognized by either party as reducible to monetised exchange, but evoked instead wider values of care, craft and community. In recognition of the more-than-monetary values being created and nurtured in repair workshops, customers sometimes sought to respond in kind by giving a gift in exchange: a pot of homemade jam, or a bunch of flowers.

Such skills, and such relationships, are endangered commodities in the twenty-first century. The past 50 years have seen the disappearance of many repair-based businesses from UK communities and high streets. The remaining businesses find that their skills and expertise are in high demand, yet many of these enterprises are run by aging proprietors with no succession plans (indeed some project participants, including Biggleston's of Hayle and Bath Typewriter Service, have closed their doors since the research began in 2010). Evidence from a related research project suggests that the regional repair industry is poorly supported and in decline, with many repairers having to supplement their income with other forms of employment (Shears, 2014). The most

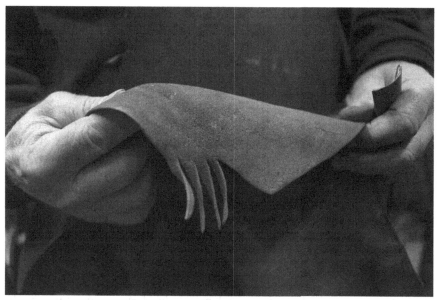

resilient businesses we worked with were family concerns, with the potential for some generational continuity, but even these establishments faced uncertain futures.

There is some evidence that the tide may be turning, with repair industries poised to benefit from drives by institutions and governments to reduce waste, lower carbon emissions and 'mainstream' sustainable development by prolonging the useful life of consumer objects and encouraging design for repairability (DEFRA, 2011a, 2011b, 2013). But there remains a deep disconnect between well-intentioned policy statements and the practicalities of economic survival for businesses like those we studied. The transition from dependence on planned obsolescence and disposability will involve promoting 'emotionally durable design' to build more meaningful and sustainable relationships between consumers and their possessions (Chapman, 2005). In this latter task, designers and consumers have much to learn from everyday menders, who are skilled in the arts of sympathetic magic required to restore valued objects to a serviceable condition, and who take pleasure in the aesthetic properties of objects as well as their transformation over time. Acts of restoration and repair are a vital part of the social lives of consumer objects. Such acts, from cleaning and maintenance to repair and restoration, change how consumers relate to objects. Repairers act, in a sense, as midwives in the birth of new relationships between people and their objects. In this way repair is a highly significant means of rekindling the aura of an object via production, rather than consumption (Gill and Lopes, 2011).

The craft skills and connected communities of making and mending enable new forms of pleasurable competence, as well as increased social wellbeing and social capital (Crawford, 2010; Gauntlett, 2011). As Gauntlett puts it: 'making *is* connecting'. Here we can substitute 'mending' for 'making' since to mend something is also to connect materials and ideas together; to connect people to other people; and, finally, to connect to menders with their wider social and physical environment (Gauntlett, 2011: p.2). It is notable how accounts of 'makers' and 'hackers' are now just as likely to focus on the places and spaces where they practice as the things they actually do (Davies, 2017). Like the workshops surveyed in this project, the 'hackerspaces' and 'makerspaces' mushrooming all over the world are often social spaces where tools, objects and makers come together in unique configurations with distinctive affordances (Kostakis, Niaros and Giotitsas, 2015). Their antecedents are to be found in the repair workshops that once were to be found all over the country, a few of which are documented in this project.

Conclusion

In recent years, much of our everyday language has become dominated by the nomenclature of neoliberal economic hegemony: a discourse of services, consumers, choice and markets, where monetary value is preeminent and social goods are subservient to the pursuit of self-interested ends (Massey, 2013). However, we found the language of value and exchange encountered in spaces of repair to be rather different to that prevalent in other areas of social and economic life. The common sense of repair shops is instead one characterised by a language of problem-solving and social exchange in which monetary interests – though certainly present – are located on the margins. Just as important in this discursive universe are expressions suggestive of the social and emotional values associated with objects, and the aesthetic and practical qualities of their material constituents and potential reparability (Spelman, 2002). The object lies at a nexus of material entanglements and social relationships, which extend between the object's custodian and its repairer.

The menders encountered in this project, together with the communities that they serve, enact key elements of a more sustainable and equitable economic framework (Gibson-Graham, Cameron and Healy, 2013). Practices of repair implicitly reframe and reclaim the economy as a space for ethical action, to be shaped for the wellbeing of individuals, communities and environments. They also prompt dialogue about how we consume, and how we can best ensure preservation of common resources. Finally, the act of repair is a crafted act of investment in the future. These repair workshops can thus be considered as important, localised 'generative spaces' for a reconfigured circular economy (Hobson, 2016). As we have shown in this chapter, one of our motivations was to reconsider and call attention to a neglected world of workplaces and practices of repair. Yet, as we discovered, many of the makers and menders we encountered are far from neglected or unappreciated by their loyal customers and the communities of which they are a part. Professional and amateur menders are eloquent and passionate about the places where they live and work, the skills and services that they perform, and the people who bring their worries and wares to them for rescue and reinvestment.

The collaborative, conversational ethos that guided our project effectively highlighted the shared social practices and pleasures that cluster around extending the useful lives of material objects. The use of photography in particular helped us to understand the affordances of objects, and the values associated with them. Treating the photograph as a crafted object in its own right – bound up in material processes of editing, printing, mounting, display, handling, wear and repair – emphasised the embedded labour in photographic practice, and opened up a sympathetic resonance with the mending and making practices we were interested in. Photographic exhibitions in public spaces provided the context for meaningful social engagement as well as the generation of valuable insights and observations that fed back into our research in various ways. Photographs acted as a catalyst for exchange between people who might not otherwise have had the opportunity to encounter each other and share their experiences. The images created the conditions of possibility for public conversations about – among other things – economic change, social resilience, sustainability and material memory. As geographies and practices of mending attract growing attention from practitioners, activists and academic researchers, it becomes ever more important to craft methods that will allow us to consider how cultures and spaces of mending are produced, and made durable.

Acknowledgements

The research discussed in this chapter was funded by the Arts and Humanities Research Council (AH/H038914/1).

Note

1 Exhibitions were held at the following times and places: June 2011, Plymouth, Devon; September 2011, Bridport, Dorset; April 2012, Exeter, Devon; and July 2012, St Austell, Cornwall. Images were also displayed at conferences, including: Mend*rs Research Symposium, July 2012, Kendal, Cumbria, and the International Visual Methods Conference, Open University, Milton Keynes, July 2011. See www.projects. exeter.ac.uk/celebrationofrepair/ for further details and a portfolio of images.

References

Adamson, G. (2007) *Thinking through Craft*. London: Bloomsbury.
Anderson, B. (2014) *Encountering Affect: Capacities, Apparatuses, Conditions*. Farnham: Ashgate.
Banks, M. (2010) Craft labour and creative industries. *International Journal of Cultural Policy*, 16(3), 305–321.

Bennett, J. (2010) *Vibrant Matter: A Political Ecology of Things*. Durham: Duke University Press.

Bond, S., DeSilvey, C. and Ryan, J.R. (2013) *Visible Mending: Everyday Repairs in the South West*. Axminster: Uniformbooks.

Boscagli, M. (2014) *Stuff Theory: Everyday Objects, Radical Materialism*. London: Bloomsbury.

Bramston, D. and Maycroft, N. (2014) Designing with Waste. In Karana, E., Pedgley, O. and Rognoli, V. (eds) *Materials Experience: Fundamentals of Materials and Design*. Oxford: Butterworth-Heinemann, 123–133.

Brook, I. (2012) Make, do, and mend: solving placelessness through embodied environmental engagement. In Brady, E. and Phemister, P. (eds) *Human-Environment Relations: Transformative Values in Theory and Practice*. London: Springer, 109–120.

Callén, B. and Criado, T. (2015) Vulnerability tests: matters of 'care for matter' in e-waste practices. *Technoscienza: Italian Journal of Science and Technology Studies*, 6(2), 17–40

Carr, C. and Gibson, C. (2016) Geographies of making. *Progress in Human Geography*, 40(3), 297–315.

Chachra, D. (2015) Why I am not a maker. *The Atlantic*. Available from: http://www.theatlantic.com/technology/archive/2015/01/why-i-am-not-a-maker/384767/ [Accessed 20 June 2016].

Chapman, J. (2005) *Emotionally Durable Design: Objects, Experiences and Empathy*. London: Earthscan.

Charny, D. (2011) *The Power of Making: The Importance of Being Skilled*. London: V & A Museum.

Chin, E. (2016) *My Life with Things: The Consumer Diaries*. London: Duke University Press.

Crang, M., Hughes, A., Gregson, N., Norris, L. and Ahamed, F. (2013) Rethinking governance and value in commodity chains through global recycling networks. *Transactions of the Institute of British Geographers*, 38(1), 12–24.

Crawford, M. (2010) *The Case for Working with your Hands*. London: Penguin.

Cooper, T. (2005) Slower consumption reflections on product life spans and the 'throwaway society'. *Journal of Industrial Ecology*, 9(1–2), 51–67.

Davies, S.R. (2017) *Hackerspaces: Making the Maker Movement*. Cambridge: Polity Press.

DEFRA (2011a) *Mainstreaming Sustainable Development: The Government's Vision and What this Means in Practice*. London: Department for Environment, Food and Rural Affairs.

DEFRA (2011b) *Guidance on Applying the Waste Hierarchy*. London: Department for Environment, Food and Rural Affairs.

DEFRA (2013) *Prevention is Better than Cure: The Role of Waste Prevention in Moving to a More Resource Efficient Economy*. London: Department for Environment, Food and Rural Affairs.

De Laet, M. and Mol, A. (2000) The Zimbabwe bush pump: mechanics of a fluid technology. *Social Studies of Science*, 30(2), 225–263.

DeSilvey, C. (2006) Observed decay: telling stories with mutable things. *Journal of Material Culture*, 11(3), 318–338.

Edensor, T. (2011) Entangled agencies, material networks and repair in a building assemblage: the mutable stone of St Ann's Church, Manchester. *Transactions of the Institute of British Geographers*, 36, 238–252.

Elster, C.H. (2003) Resistentialism. *The New York Times Magazine.* Available from: http://nyti.ms/165ESdG [Accessed 11 February 2017].

Fickey, A. (2011) 'The focus has to be on helping people make a living': exploring diverse economies and alternative economic spaces. *Geography Compass*, 5(5), 237–248.

Gauntlett, D. (2011) *Making is Connecting: The Social Meaning of Creativity, from DIY and Knitting to YouTube and Web 2.0.* Cambridge: Polity.

Gibson-Graham, J.K., Cameron, J. and Healy, S. (2013) *Take Back the Economy: An Ethical Guide for Transforming our Communities.* Minnesota: University of Minnesota Press.

Gill, A. and Lopes, A. Mellick (2011) Recoding abandoned products: student visual designers experiment to sustain product lives and values. Proceedings of the International Conference 2011 of the DRS Special Interest Group on Experiential Knowledge, EKSIG 2011: SkinDeep: Experiential Knowledge & Multi Sensory Communication, 23–24 June 2011, the University for the Creative Arts, Farnham, Surrey, UK. Available from http://www.experientialknowledge.org.uk/conference_2011.html [Accessed 11 February 2017].

Graham, S. and Thrift, N. (2007) Out of order: understanding repair and maintenance. *Theory, Culture and Society*, 24(3), 1–25.

Gregson, N., Crang, M., Ahamed, F., Akhter, N. and Ferdous, R. (2010) Following things of rubbish value: end-of-life ships, 'chock-chocky' furniture and the Bangladeshi middle class consumer. *Geoforum*, 41(6), 846–854.

Gregson, N., Metcalfe, A. and Crewe, L. (2007) Identity, mobility, and the throwaway society. *Environment and Planning D: Society and Space*, 25(4), 682–700.

Gregson, N., Metcalfe, A. and Crewe, L. (2009) Practices of object maintenance and repair: how consumers attend to consumer objects within the home. *Journal of Consumer Culture*, 9(2), 248–272.

Hackney, F. (2013) Quiet activism and the new amateur: the power of home and hobby crafts. *Design and Culture*, 5(2), 169–193.

Hawkins, H. (2014) *For Creative Geographies: Geography, Visual Arts and the Making of Worlds.* Routledge: Abingdon.

Highmore, B. (2016) Capacious aesthetics (review essay). *New Formations*, 89/90, 234–242.

Hitchings, R. (2006) Expertise and inability: Cultured materials and the reason for some retreating lawns in London. *Journal of Material Culture*, 11(3), 364–381.

Hobson, K. (2016) Closing the loop or squaring the circle? Locating generative spaces for the circular economy. *Progress in Human Geography*, 40(1), 88–104.

Houston, L. (2017) The timeliness of repair. *Continent*, 6(1), 51–55.

Ingold, T. (2009) The textility of making. *Cambridge Journal of Economics*, 34(1), 91–102.

Ingold, T. (2013) *Making: Anthropology, Archaeology, Art and Architecture*, Abingdon: Routledge.

Jackson, S.J. (2014) Rethinking repair. In Gillespie, T., Boczkowski, P.J. and Foot, K. A. (eds) *Media Technologies: Essays on Communication, Materiality, and Society.* Cambridge: MIT Press, 221–240.

Janigo, K., Wu, J. and DeLong, M. (2017) Redesigning fashion: an analysis and categorization of women's clothing upcycling behaviour. *Fashion Practice: The Journal of Design, Creative Process and the Fashion Industry*, 9(2), 254–279.

Jennings, P. (1950) *Oddly Enough.* London: Reinhardt and Evans.

Keulemans, G. (2016) The Geo-cultural Conditions of *Kintsugi*, *The Journal of Modern Craft*, 9(1), 15–34.

Kopytoff, I. (1986) The cultural biography of things: commoditization as process. In Appadurai, A. (ed.) *The Social Life of Things: Commodities in Cultural Perspective.* Cambridge: Cambridge University Press, 64–94.

Kostakis, V., Niaros, V. and Giotitsas, C. (2015) Production and governance in hackerspaces: a manifestation of Commons-based peer production in the physical realm? *International Journal of Cultural Studies*, 18(5), 555–573.

Lepawsky, J., Liboiron, M., Keeling, A. and Mather, C. (2017) Repair-scapes. *Continent*, 6(1), 56–61.

Luckman, S. (2012) *Locating Cultural Work: The Politics and Poetics of Rural, Regional and Remote Creativity.* London: Palgrave Macmillan.

Luckman, S. (2015) *Craft and the Creative Economy.* New York: Springer.

Massey, D. (2013) Vocabularies of the economy. *Soundings*, 54, 9–22. Available from: https://www.lwbooks.co.uk/sites/default/files/s54_02massey.pdf [Accessed 28 March 2017].

Morozov, E. (2014) Making it: pick up a spot welder and join the revolution. *The New Yorker.* Available from: http://www.newyorker.com/magazine/2014/01/13/making-it-2 [Accessed 6 July 2016].

Napier, A.D. (2014) *Making Things Better: A Workbook on Ritual, Cultural Values, and Environmental Behaviour.* Oxford: Oxford University Press.

Orton-Johnson, K. (2014) DIY citizenship, critical making, and community. In Ratto, M. and Boler, M. (eds) *DIY Citizenship: Critical Making and Social Media.* Cambridge: MIT Press, 141–156.

Paton, D. A. (2013) The quarry as sculpture: the place of making. *Environment and Planning A*, 45(5), 1070–1086.

Pink, S. (2009) *Doing Sensory Ethnography.* London: Sage.

Pink, S. (2013) Prologue: repairing as making. In Bond, S., DeSilvey, C. and Ryan, J.R., *Visible Mending: Everyday Repairs in the South West.* Axminster: Uniformbooks, 13–14.

Rose, G. and Tolia-Kelly, D. (2012) Visuality/materiality: introducing a manifesto for practice. In *Visuality/Materiality.* Farnham: Ashgate.

Russell, A. and Vinsel, L. (2017) Let's get excited about maintenance! *New York Times.* Available from: https://www.nytimes.com/2017/07/22/opinion/sunday/lets-get-excited-about-maintenance.html [Accessed 15 August 2017].

Schumacher, E.F. (1973) *Small is Beautiful: A Study of Economics as if People Mattered.* New York: Harper and Row.

Sennett, R. (2009) *The Craftsman.* New Haven, CT: Yale University Press.

Shears, J. E. (2014) *Prevention is Better than Cure, but is there the Capacity for it to Succeed?* MSc Dissertation, University of Exeter, Penryn Campus.

Spelman, E.V. (2002) *Repair: The Impulse to Restore in a Fragile World.* Boston: Beacon Press.

Thomas, N.J., Harvey, D.C. and Hawkins, H. (2013) Crafting the region: creative industries and practices of regional space. *Regional Studies*, 47(1), 75–88.

Thompson, M. (1979) *Rubbish Theory: The Creation and Destruction of Value.* Oxford: Oxford University Press.

Van Nes, N. (2010) Understanding replacement behaviour and exploring design solutions. In Cooper, T. (ed.) *Longer Lasting Products: Alternatives to the Throwaway Society.* London: Gower, 107–131.

13 Re-lighting the Castle Argyle

Making, restoration, and the biography of an immobile thing

Dydia DeLyser

One evening in fall of 2013, at a building in Hollywood, California, in a small ceremony for City Council representatives, local historians, and the building's owners and residents, a neon-sign craftsperson threw a switch to light a large roof-top neon sign (Figure 13.1).

The sign, which read "Castle Argyle" had never fully left the rooftop where it stood, but it had already had a long and complex history: it had been erected when the building was new in 1930, had gone dark during WWII, had been modernized and relit after the War, had decayed and gone dark again, had been patched up and relit in the 1990s, had gone dark once again, and now had been restored as closely as possible to its original appearance, and relit once more. In this chapter I use interviews, observation, and archival research to uncover this history, and, in so doing, to understand the process of the sign's restoration and re-making. I trace the biography of one thing that scarcely moves, and follow the process of its restoration, to draw the materially engaged and embodied practices of restoration into discussions of making.[1]

By "restoration" I mean efforts to return a thing (any thing – a building, an artwork, an automobile, or a neon sign) to something close (often as near as possible) to its supposed original state. Efforts at restoration include a diverse array of practices and processes, and restoration being practiced in different ways, for different things, in different historical periods. It is always in dispute, and debates have always swirled around the permissibility or impermissibility of modern modifications, replacement of original parts/fabric/elements with new materials or replicas, and indeed around whether restoration should be carried out at all. Thus, for different things, public and professional ideas about what constitutes restoration are constantly changing (Lowenthal 2016). Key here, in the context of practices of making, is that restoration departs significantly from maintenance and repair: both maintenance and repair are ordinary and expected undertakings that happen to nearly all things (particularly those mechanical). Where maintenance involves efforts to ensure the continued smooth operation of a thing, and repair involves returning a thing to smooth operation after it has failed, restoration goes much further: restoration involves returning a thing to its (supposed) original made state, making something old like new again. The tools, techniques, skills, and

Figure 13.1 The "Castle Argyle" sign after restoration in 2013
Photograph by Dydia DeLyser

finances are different. But like maintenance and repair (Graham and Thrift 2007; Gregson et al. 2009), restoration is also a skilled practice of making (DeLyser and Greenstein 2015a, 2017).

The growing literature in geography and other disciplines on practices of making has tended to focus on creative and home- or small-workshop-based handcraft practices that make things from scratch, or remake and repair things that are old – like crafting from yarn, caning chairs, custom surfboard making, or clock repair (DeSilvey, et al. 2013; Bond et al. 2013; Thomas et al. 2013; Warren and Gibson 2014). Typically, as is true with the examples above, the made and remade things under study are comparatively small – they can fit easily into a small craft studio. In this chapter I shift our attention to things that are very large. The thing at the heart of this chapter is, though not as large as a building (Jacobs 2006; Rose et al. 2010), nevertheless far too large for quaint crafting. But size, of course, is seldom the main point. So in this chapter I also draw out another difference: the practices of making demanded by restoration are distinct. As others have shown, the practices of making from new and making from old are unalike – compared to making from new, the skilled, embodied practices of maintenance and repair typically involve different craftspeople, different skills, and different engagements with materials and materiality (Graham and Thrift 2007; Gregson et al. 2009; Edensor 2011; Strebel 2011; Bond et al. 2013; DeLyser and Greenstein 2015a; DeLyser and Greenstein 2017). Here I seek to extend those distinctions to the practices of restoration (see also DeLyser and Greenstein 2015a, 2017).

At the same time, in follow-the-things literature, and in parallel work on commodity chains, scholars have focused on things with mobile biographies – tracing the geographies of things through production, use, repurposing, ridding, and re-use again (Cook 2004; Cook et al. 2006; Foster 2006; Gregson et al. 2007; Ramsay 2009; Gregson et al. 2010; Cook and Woodyer 2011; Lepawsky and Billah 2011; Lepawsky and Mather 2011; Pfaff 2010; Crang et al. 2013; DeLyser and Greenstein 2015a; DeLyser and Greenstein 2017). In this chapter, I focus on a thing that has essentially not moved in the roughly

90 years since it was erected in the place where it still resides (on immobile things see also Jacobs 2006 and Edensor 2011). It is a thing that has nevertheless been made, re-made, and re-re-made over that time span. In this chapter I thus seek to complicate the mobile biographies of things that trace a life cycle from production or making through consumption and disposal across different spaces of use, reuse, and disuse.

Here, I join other recent scholars of making to present an ethnographic account of making's embodied practices, in this case the restoration of one large roof-top neon sign – that of the Castle Argyle Apartments in Hollywood, California, originally built in 1930[2] and restored most recently in 2013 – to show how restoration, as a practice of making, requires significant unmaking and remaking, just to return to the original "made" state. I reveal how, more than other forms of making, the practices of restoration must bring together embodied expertise with historical knowledge and ingenuity. In fact, the practices of re-making and re-lighting a neon sign like this one ignite its biography anew. But unlike other commodities, such large signs are hardly mobile. So this chapter follows a thing that really does not move to show how, because it must remain anchored in place, its very geographical fixity can inspire a new mobility, and new community value around the restored or re-made object. Making, thus, is a practice at once cultural and political.

Neon signs in the US

Beginning in the late nineteenth century, roughly contemporaneously with electric street lighting and indoor electric lighting, electric signs became wildly popular in American cities ("Fifty Years of Electric Signs" 1956; Bowers 1998; Dillon 2002; Brox 2011; Ribbat 2013; Rinaldi 2013). Large and bright and transforming the night, these signs advertised products in urban commercial districts (like New York's famous Times Square), but also business at their place of business (Treu 2012; Rinaldi 2013; DeLyser and Greenstein 2015b). Originally, electric signs were lit (inside and/or outside) by incandescent bulbs, and often they were large, sometimes very large, requiring dozens, hundreds, or even thousands of bulbs. Such enormous bright signs drew a great deal of attention but also required a great deal of maintenance – the renowned Hollywood sign, for example, built in 1923 and lit originally with 4,000 bulbs, demanded one person's full-time attention just to service the bulbs (DeLyser and Greenstein 2015b; see also "Fifty Years of Electric Signs" 1956; Rinaldi 2013).

By the mid-1920s a new lighting technology arrived in the US from France. Then called "luminous tubing" and today known simply as "neon" (for one of the gases used inside the tubing) this new lighting was nothing short of revolutionary. Where incandescent bulbs formed letters and shapes out of dots (bulbs), luminous tubing could be bent by skilled craftspeople into any shape or word forming bands of continuous light. Where bulbs eventually came in several colors, luminous tubing evolved into a virtually unlimited range of shades and hues. And, where bulbs were costly to operate and maintain, luminous tubing was cheap to run and required little or no maintenance; it was ideal for those places difficult to reach, those high up on buildings, on roofs, or those – like the Hollywood sign – on mountain tops.[3] By the mid-1930s neon became the dominant form of illumination in lit signs across the US (see also "Fifty Years of Electric Signs" 1956; Rinaldi 2013; DeLyser and Greenstein 2015b).

From the 1920s to the 1940s large roof-top neon signs proliferated in cities across the US. Installed many stories up they were visible from a great distance. Often they advertised local businesses – car dealers, hardware stores, hotels or apartment buildings, and were built on the roofs of the businesses they advertised. Though much outdoor advertising (think of billboards (see Gudis 2004)) can be separated from the businesses the signs advertise, these large rooftop signs have nearly always been anchored to place, built onto the very businesses they promote. Thus the signs, the buildings, and the businesses developed their public identities in sync, and each sign was unique – each a work of art/design and skilled craftspersonship (Treu 2012).

During WWII however, signs – including of course roof-top signs – across the US were blacked out, as the West Coast in particular was sensitive to threats of enemy attack (Starr 1999; DeLyser and Greenstein 2015b). After the war, and after years unlit, many of these signs were not relit. In other cases unrelated to the War, when the businesses closed or changed hands the signs were turned off. In either case, the darkened signs were often not removed: the signs themselves and their steel frameworks are heavy and expensive to tear apart and haul off; old owners didn't want to invest money repairing an old sign dark for years; new owners simply ignored the darkened signs. In either case, their durable construction meant that signs remained standing despite their increasing age and exposure to the elements.

By the 1960s when landscape aesthetics had changed and signs of all kinds became considered an intrusive and largely undesirable element of urban advertising, "Scrap Old Signs" campaigns tore down as "blight" disused signs across the US (Gudis 2004; Rinaldi 2013). Neon became yesterday's lighting, the tawdry signs associated with low-rent commercial districts and their seedy businesses. But again, many rooftop signs remained due to the cost of removing them. As fate often has it, not long after the old signs were disdained and/or removed, those same old signs began to be valued: the late 1970s and early 1980s saw a revival of neon and a gradual resurgence of interest in old signs along with it (Rinaldi 2013; DeLyser and Greenstein 2015b).

By the late 1990s, though aesthetics about "sign clutter" and morality-based assessments of (neon in) advertising had not changed, surviving old signs became considered "collectible" and "antiques"; they had become valuable markers of personal and community identity (Rinaldi 2013; DeLyser and Greenstein 2015b). As a result, across the US municipalities, building owners, community organizations, and historical societies began raising money to restore the remaining signs, even when the businesses named were no longer extant (see, for example, Starr 1999).

Los Angeles, at the turn of the twenty-first century, became home to the most significant example of urban municipal sign-restoration efforts in the US: the LUMENS project (with LUMENS standing for Living Urban Museum of Electric and Neon Signs) of the late-1990s and early 2000s brought together the City's Department of Cultural Affairs and the non-profit Museum of Neon Art with the support of the Mayor and City Council to focus on the city's large remaining cache of original, but by then long-unlit, incandescent and neon signs. The project, which historian Kevin Starr termed "one of the most imaginative and cost-effective redevelopment schemes in Los Angeles history", a project that, he wrote, "literally relit history and ... brought forward to present-day L.A. the mood and mystery of the city in its [Raymond] Chandler years."[4] LUMENS eventually relit some 80 or more signs in Los Angeles's Downtown, its Wilshire corridor, and in Hollywood (Starr 1999).[5]

Unfortunately, while the project paid for these signs' original relighting, the project favoured cost-effectiveness and therefore repair not careful restoration, and no funds were set aside for maintenance or future repair. Further, with comparatively easy-to-get city money available for the initial projects, some of the relit signs suffered from poor workpersonship. Soon, many went dark again. In some cases, like that of the 1923 Jensen's Recreation Center sign (today probably the oldest remaining light-bulb spectacular[6] in the US), a local organization – the Echo Park Historical Society – funded the re-relighting just a decade after the original LUMENS work had been done; the sign had been relit after some 80 years; it needed re-relighting after just ten more. In other cases private businesses and building owners now found their old roof-top signs of such significant importance that they paid to re-relight the signs themselves (this eventually happened at Jensen's too: less than ten years after the second relighting decay and exposure made the sign dark again, and the new building owners agreed to fund the costly restoration the sign had needed all along). Privately funded restoration and relighting was the case for Castle Argyle, the subject of the rest of this chapter.

The Castle Argyle

The property now known as the Castle Argyle, located in Hollywood, California on Argyle Avenue just north of Franklin Avenue, was originally developed by Dr. Alfred Schloesser (who changed his name to Castle during WWI), for himself, his family, and their lavish entertaining. Wealthy from Nevada silver mines, the Schloessers had originally built the so-called "Glengarry Castle" – modeled on Emma Schloesser's ancestral home in Scotland – around the turn of the twentieth century, but by 1912 found it too small for their growing family and expansive entertaining practices, and so built the larger Castle Sans Souci (Figure 13.2) on a lot across the street (Rasmussen 1996). In 1928,

Figure 13.2 The "Castle Sans Souci," home of the Schloesser family. Postcard ca. 1906. Collection of Paul Greenstein
Photograph by Dydia DeLyser

with more development in surrounding Hollywood, they demolished their second castle in favor of a speculative investment in the seven-storey "Castle Argyle Arms" apartments. Designed as luxury flats the Argyle became a chic place to stay, favored by stars like Clark Gable, Howard Hughes, and even Ronald Regan (Chazanov 1993; Rasmussen 1996). Like many other luxury buildings then being built in Hollywood, by 1930 the new and modern Castle Argyle Arms sported a large neon sign advertising the property on its roof.[7]

However, by the 1940s, when the Hollywood Freeway was cut through the mountains and properties nearby, several years of profoundly disruptive construction transformed the neighborhood; the wealthy moved away, the building fell on harder times (see Miller 2009). The Castle Argyle was remodeled to make the apartments considerably smaller and more modest, and in the 1950s purchased by Southern California Presbyterian Homes (renamed be. group in 2011) who today continue to provide affordable housing chiefly for senior citizens.[8] In 2012 it was this organization that sought to restore and re-re-light the sign, contracting with neon-restoration specialist Paul Greenstein to do the work. But restoration of the sign was not their original plan; their initial focus was on repair – enough repair to get the sign lit. This is not unusual for processes of restoration: while sometimes complete restoration is proposed upon a project's undertaking, other times it is a commitment entered into only once repair work has been initiated and found – for any of various reasons (financial, person hours, future projected repair needs, etc.) – to be infeasible or inadvisable.

The Argyle sign, though one of the older still-extant illuminated rooftop signs in Los Angeles, was built in the fashion still standard for rooftop electric signs (indeed, though neon is some 90 years old in the US, the practices and processes of its making have changed little, and it remains a skilled handcraft industry): for Argyle a large and heavy-gauge framework of horizontal, vertical, and diagonal angle iron provided bracing and structural support for the sign's letters, and was itself fastened to the building's roof. The framework's horizontal members provided orientation for the sign's letters, each of which was made from light-gauge sheet steel. These letters, called "channel letters" had fronts and backs in the shape of the letters themselves, and sides that served to project the letters away from the framework. Thus each four-foot-tall letter had a three-dimensionality that was visible even from below.

Mounted to the front (or outside) surface of each letter was the luminous tubing: tubes of glass, bent by hand in the exact shapes of the letters, and held just inches away from the steel by means of glass tube supports. The glass tubing for each letter was formed in one continuous shape, with the ends often coming close together but never connecting. Inside the tubing was a gas (neon, or argon mixed with mercury) that, when subjected to high voltage, would emit a characteristic colored glow. At the precisely positioned places on the fronts of the letters where each end of the glass tubing stopped, a circular hole punched in the steel allowed the tubing to make a right-angle bend and disappear into the letters (Figure 13.3). Here, behind the visible surface of each letter, each end of the tubing was fused to an electrode, with wires protruding. Those wires were then spliced to high-tension wire, and connected to four 12,000 volt transformers that stepped up the ordinary 110-volt electrical current of the building to the nearly 50,000 volts needed to electrify the gas inside the tubing.[9] The wiring, as was/is typical, was protectively housed in steel conduit, and connected to the building's power down below. A switch on a timer could control the sign so it would illuminate only during peak hours of the night. All of these aspects of the Argyle sign were typical and ordinary, just like hundreds (and once thousands) of other signs across the US. But, as with any restoration (or re-making) project, the actual specifics of this sign were unique.

The years had not been kind to the Castle Argyle sign, and the LUMENS-sponsored re-lighting had not, in fact, been a restoration but rather what Greenstein termed a "patch job": Greenstein learned when he got up on the roof that the sign had not been significantly re-wired, or even repainted. Under LUMENS they had done only enough work to repair and light the glass. By 2012, when Greenstein inspected the sign for the first time, he found that the paint was peeling, "the transformers were burned out, the metal was rotted away, half of the glass was broken, all of the glass was the wrong pattern. Nothing worked." It was, he said, "skanky, scabberous, beat up, and hammered."

The sign was in such poor condition that Greenstein realized it needed more than repair and maintenance. In order for Castle Argyle to become a

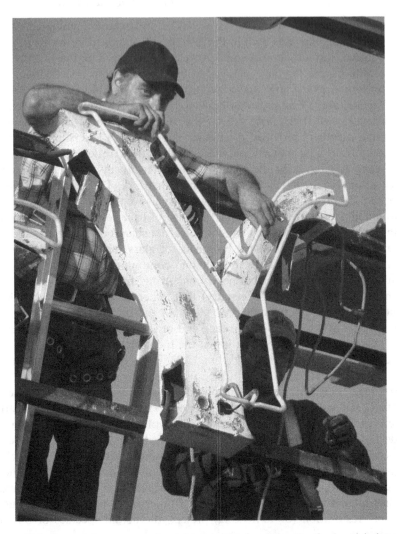

Figure 13.3 Greenstein removes the tubing intact from the Castle Argyle's letter Y. The white tubing is bent in one continuous shape; each end, before its removal, disappears through round holes visible in the steel "can" at the letter's base. Also visible here is the original green paint (at left) and a good deal of rust where the newer white paint has peeled away, and two open access panels (one at top left, the other at bottom)—these had allowed water and birds to get inside the letters

Photograph by Dydia DeLyser

working and yet low-maintenance sign the building's management would have to pursue not another patch job, but a complete restoration. At some point, he argued, repair leads simply to more repair; only restoration could stop the cycle. The management agreed, and so a repair project became a restoration project – the practices of making shifted significantly from those of returning something that was broken to working status, where the repairs would likely be visible (see Bond et al. 2013), to those of making something old new again, with the labors of restoration not to be seen in the completed, working project.

And so work commenced. The first task, while the un-restored letters were still in place, was rewiring, this time, from the fuse box on the ground floor all the way to the letters on the roof (over ten stories up). Greenstein and his electrician installed four new 12,000-volt transformers in the exact locations where the old ones had been, and were careful to run new conduit along exactly the same route where the old (now rusted and rotten) conduit had been. The old conduit and the old sign served as a pattern for the restoration.

Then Greenstein and his crew of two removed the glass tubing from each letter. Because some of it came off intact, that glass could be electrified to reveal that its original color had been white. But that color signaled to Greenstein that these glass letters were not the sign's original glass letters: he knew that there was no white neon in 1930. It was the first part of a puzzle in the sign's restoration, and it took a skilled maker to understand its meaning.

Next the crew set about to remove the metal letters from the sign's frame-work so the dilapidated letters could each be restored. The nuts holding the letters to the sign's steel frame had not been removed since 1930. They had long-since rusted into place, so now they had to be ground and cut away from the frame (Figure 13.4). Once freed, the letters could be carefully lowered, by rope, to the rooftop below. Inspecting them immediately revealed surprises and set Greenstein to try to understand what the sign had originally looked like – a key part of any restoration process.

The front faces were made of delicately corrugated galvanized steel collo-quially known today as "ripple tin," – this was common in signs of the 1920s but disused by the late 1930s, so the letters were all clearly original. The ripple tin, because it was galvanized, remained in good condition, but the steel sides and backs of the letters had not been galvanized and so were rusted, and, in places, completely rotted away. In addition, the Argyle sign had been subject to one of the chief perils for outdoor signs: over the years, holes and open access panels had allowed water and pigeons in – now birds' nests, dung, and (electrocuted) dead bird bodies filled many of the letters, the acid eating away at the metal even further. Greenstein would have to carry the letters down off the roof, repair each one by hand, and make new hand-cut sides and backs for several of them. Each letter would demand individual attention and treatment – in re-making and restoration, unlike in making from scratch, no blanket approach is possible.

When Greenstein first took the letters down from the sign, the paint remaining on them was white (see Figure 13.3). But the peeling white paint

Figure 13.4 To free the letters Greenstein grinds away the nuts that had held them in
place since the sign was built in 1930
Photograph by Dydia DeLyser

revealed the original dark green paint beneath – the deepest, oldest layer of
paint was dark green, so the sign had originally been dark green. Further
evidence of what the letters originally looked like was revealed when Greenstein had the letters chemically stripped. Because over time paint actually
etches metal, on old signs that have had paint for many decades, that paint
leaves shadows of itself even after it is removed. Tracing the shadows then
reveals areas that once held different colors of paint. In Argyle's case, stripping
the paint revealed the shadows of a pinstripe along the edges of each letter,
but not, of course, the pinstripe's original color. Familiar with pinstriping
techniques and color choices in Argyle's era, Greenstein knew that to highlight the dark green letters the pinstripe would have been a lighter color. With
no evidence of the color remaining anywhere on the sign, he simply had to
select a color – he decided on light blue: not as strong a contrast as white,
and, particularly against a blue sky, complementary to the dark green.

Also original when the letters came down were some of the tube supports
that held the glass to (and insulated it from) the steel letters, and these
answered other questions about what the sign had originally looked like.
Some of them, Greenstein observed, were in the wrong places for the white
glass letters he had just removed: There were screw-holes for tube supports
that did not match with the glass. There were extra holes in the sign that were

no longer used. Some of the original glass tube supports were still fastened to the sign (a few had even turned purple from 80 years of sun exposure), though they were no longer supporting tubes. But why?

The white tubing of the restored sign was "two-stroke neon," meaning that each letter, when lit, was illuminated by two glowing parallel lines of glass that traced the outline of the letter's shape (see Figure 13.3). Had the Argyle sign originally held three-stroke neon, with a third tube filling the center of each letter? By carefully tracing, based on the patterns of holes in the letters, where the glass letters would have had to have run, Greenstein determined that it was not possible that the sign had been three-stroke neon: the holes and the housings did not match. He realized instead that it had been a single-tube sign that was changed to a double-tube sign some time, years after it was first made.

What Greenstein knew, based on years of experience making and restoring neon signs, was that in the 1940s, when the Castle Argyle was remodeled, neon was undergoing a transformation – designers were adding more glass, more tubes to their letters, to make signs brighter and flashier. But in the 1920s, neon had not become garish yet – that was a developmental stage that came later. So though the glass he removed from the sign was double-stroked, he realized that the original sign had had single-stroke tubing and that was how he wanted to restore it.

It wasn't easy to convince the Argyle management though: neon's history has been ill-documented and little written about; much of what has been written relies upon myth rather than documentation (DeLyser 2014; but see Rinaldi 2013). Most Americans today understand neon's history instead by what they remember about it, and what is visible in films of the mid-twentieth century. Since most signs from the 1920s had vanished before current living memory, what most people now remember is neon signs from the post-War period. And that corresponds to popular cinema's representations of neon in that same period: most film audiences today are unfamiliar with silent-film and early talkie representations of neon in the 1920s and 1930s – glowing testimony to an optimistic time before Depression and War, radiant display of unquestioned technological prowess – but are well-familiar with post-War noir representations of seedy neighborhoods illuminated by garish, flashing signs (Rinaldi 2013; Ribbat 2013).

To return the sign from double-stroke neon to single-stroke neon would mean a "less is more" approach – less glass, less light, but more elegance and the original look of the period when the sign was built. To convince the building's management of this approach Greenstein asked a friend from Hollywood Heritage Association to draw a digital mock-up of what the sign would look likr in single-tube neon – that convinced them.

The next decision was about color, since the original colors for the tubing were lost. Greenstein knew that there were only three colors of neon in 1930: blue, red/orange, and rarely used green. He chose blue letters with a red accent for elegance and because of the way green or red would have looked with the dark green steel letters (Figure 13.5)

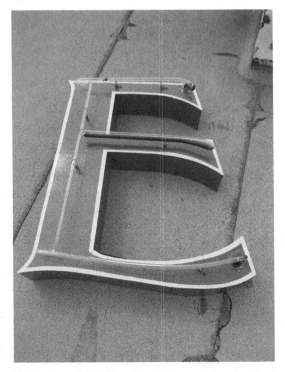

Figure 13.5 Castle Argyle's first E, after restoration. The "ripple metal" is visible on
the letter's front, as are the repainted original green color, the pale blue
pinstripe, and the newly bent, single-stroke neon tubing
Photograph by Dydia DeLyser

   With color choices made, Greenstein drew new patterns for the glass and
had a skilled neon tube bender bend new letters to match the sign. The
bender, working with the same techniques and essentially the same equipment
neon craftspeople have been using for some one hundred years, bent each
piece of tubing by hand over special gas fires, matching the softened glass to
Greenstein's pattern. Once the letters had been bent, the bender next sealed
the tubes by splicing glass-encased electrodes to either end and then evacuated the air from them first by pumping the air out to approximate a
vacuum, and then by bombarding the sealed tubes with so much electricity
that any molecules remaining effectively burned away. Next he introduced a
closely measured amount of a noble gas into the evacuated letters: pure neon
for the red, and argon with a drop of mercury added for the blue. Finally, he
tested the new letters by "burning them in" for several hours at increased
voltage. Only then were they ready to be mounted to the steel letters of the
sign. The restored sign, even in the details of how the neon tubing was fabricated, underwent practices of making almost identical to those of its original
manufacture – the embodied practices of making neon have changed little in

one hundred years, and so the practice of restoring neon signs has remained similarly constant.

Next Greenstein used copper wire to mount the glass letters to the glass tube supports on the repaired, repainted letters: each letter was fully re-assembled on the ground and ready to light, even before being returned to the rooftop and the sign framework (see Figure 13.5). Then on one day in fall of 2013 Greenstein and his crew brought the restored letters to the rooftop and, letter by letter, re-assembled and re-connected the Castle Argyle sign (Figure 13.6, Figure 13.7). In November, at a ceremony in the presence of the City Council

Figure 13.6 The newly restored Castle Argyle letters are re-installed in their original locations on the rooftop frame
Photograph by Dydia DeLyser

Figure 13.7 The restored Castle Argyle by day
Photograph by Dydia DeLyser

representative and Argyle executives and residents, Greenstein threw the switch to relight the sign. (see Figure 13.1) For the first time in 50 years, Argyle was its old self again.

Conclusion

Much of the scholarly work on making has focused on production, on objects made new. But, as scholars are increasingly pointing out, repair, maintenance, and restoration are significant and encompass distinctly different practices of making (Graham and Thrift 2007; Gregson et al. 2009; Edensor 2011; Strebel 2011; Bond et al. 2013; DeLyser and Greenstein 2015a; DeLyser and Greenstein 2017). Indeed, as Josh Lepawsky and Mostaem Billah (2011: 122) suggest, "at least as important [as production] are the practices that capture and create value anew," those practices that sustain and restore the thingness of our things (see also Gregson et al. 2010). This is because, as Stephen Graham and Nigel Thrift (2007: 4) observe, repair and maintenance – and, I would add, restoration – draw our focus to "the importance of human [handcraft] labour and ingenuity" as part of the essential practices of making. This chapter's focus specifically on restoration also demonstrates the importance, as Nicky Gregson, Alan Metcalfe, and Louise Crewe (2012) have noted, of first the "after lives" of made things – lives beyond their original production and consumption – and, second, the social embeddedness of those afterlives – in this case the Argyle sign long stood in darkness and could be restored only once its community value had been established for the third time.

In order to draw forward this work on making and maintenance by focusing on the skilled practices of restoration, I have historically and ethnographically followed a thing that basically did not move. But most of the works that have followed things – from, for example, Julia Pfaff's (2010) work following one mobile phone in Africa, to Nicky Gregson and colleague's work (Gregson et al. 2010) tracing the end-of-life repurposings of cargo ships – have traced the biographies of highly mobile commodities. In Argyle's case, the entirety of the thing really did not and *could not* move (the steel framework has always stayed precisely in place; the letters had to be returned to the rooftop once they were restored). Immobile things like the Castle Argyle sign, then, have distinct geographies. Instead of being mobile, an immobile thing "construct[s] a mobility around itself," (Davidson 2012: 479) drawing in the movements of people, ideas, and things – in this case, the people, ideas, and things required for its restoration. By so doing, today, the Castle Argyle sign remains standing, now re-lit, on the roof where, in 1930, it began.

In this chapter I have endeavored to build from Ian Cook and Tara Woodyer's (2011: 231) call for "more empirical encounters that include vivid portrayals of the affection, passion and creativity pervading our everyday relations with things" in an emotionally engaged geography that is imaginative, politically informed, and critically positive. To do so I have traced the biography and restoration of one large roof-top neon sign – the Castle Argyle's – to reveal the ways that "making" can involve significant "unmaking" and "re-making" just in order to return to the original "made" state, attempting to draw attention to the embodied practices of restoration work, and the ways that that work must combine historical expertise, embodied skill, and ingenuity. That combination does not come without both effort and emotion (DeLyser and Greenstein 2015a, 2017); as I have shown, the labor of restoration, as a practice of making, involves not just skill but care.

Notes

1 This chapter is part of a much larger project on the historical geography of the neon-sign industry in the US. In this chapter, I use one sign project, and highlight the efforts of just one person, Paul Greenstein, the neon maker who restored this sign. Though others were essential to this project as well, for reasons of institutional ethics, only Greenstein is described by his real name; the identities of the others remain protected. The larger project includes dozens of interviews with neon craftspeople across the US, and observations at neon shops and sign-installations sites.

2 The City of Los Angeles Department of Building and Safety permitted an "all metal roof sign" for the building on 23 September 1930; LADBS document number 1930LS22894, available at LADBS.

3 To be clear, the Hollywood sign was never illuminated by neon tubing. Had it been made a few years later, however, it likely would have been – and that would have saved a great deal of maintenance.

4 See http://www.neonmona.org/mona_news/LUMENS%20one%20page%209-29-08.
   pdf last accessed 2 January 2015.
5 See also: http://www.welcometolace.org/events/view/portable-city-projects-al-noda
   l-lumens-project/ last accessed 2 January 2015.
6 A "spectacular" is a sign that flashes in such a way that it suggests movement. In
   the case of Jensen's Recreation Center, not only did the words of the name flash on
   in sequence, but a bowler could be seen releasing a ball toward a set of pins, the ball
   then tracks – flash by flash – towards the pins, until it contacts, and appears to
   knock over the pins, in a shower of flashing lights.
7 The City of Los Angeles Department of Building and Safety document number
   1930LS22894; permitted on 23 September 1930; LADBS. See note 2.
8 See http://www.thebegroup.org/about-us last accessed 2 January 2016.
9 Though neon signs require high voltage, their low amperage requirements mean
   that neon signs use very little electricity. This was one of neon's early advantages
   over incandescent signs, and remains an advantage today – though today's LED
   (light-emitting diode) signs are also energy efficient, they cannot compete with neon
   for energy efficiency, longevity, and renewability.

References

Anderson, C. 2012. *Makers: The New Industrial Revolution*. New York: Crown
   Business.
Anderson, J. 2012. Heritage discourse and the desexualization of public space: The
   "historical restorations" of Bloomsbury's squares. *Antipode*, 44(4): 1081–1098.
Bond, S., DeSilvey, C. and Ryan, J.R. 2013. *Visible Mending: Everyday Repairs in the
   South West*. Axminster: Uniformbooks.
Bowers, D. 1998. *Lengthening the Day: A History of Lighting Technology*. New York:
   Oxford University Press.
Braudy, L. 2012. *The Hollywood Sign: Fantasy and Reality of an American Icon*. New
   Haven, CT: Yale University Press.
Brox, J. 2011. *Brilliant. The Evolution of Artificial Light*. New York: Mariner
   Books.
Carr, C. and Gibson, C. 2015. Geographies of making: Rethinking materials and skills
   for volatile futures. *Progress in Human Geography*, 1–19. doi:10.1177/
   0309132515578775.
Charney, D., ed. 2011a. *The Power of Making*. London: V&A Publishing.
Charney, D. 2011b. Thinking of making. In *The Power of Making*, ed. D. Charney.
   London: V&A Publishing, 6–10.
Chazanov, M. 1993. Renters seek to buy piece of history, housing. *The Los Angeles
   Times*. 15 July. Available at: http://articles.latimes.com/1993-07-15/news/we-13151_
   1_castle-argyle last accessed 2 January 2016.
Cook, I. 2004. Follow the thing: papaya. *Antipode*, 36: 642–664.
Cook, I. et al. 2006. Geographies of food: following. *Progress in Human Geography*,
   30(5): 655–666.
Cook, I. and Tolia-Kelly, D. 2010. Material geographies. In *Oxford Handbook of
   Material Culture Studies*, eds. D. Hicks and M. Beaudry. 99–122. Oxford: Oxford
   University Press.
Cook, I. and Woodyer, T. 2011. Lives of things. In *The New Companion to Economic
   Geography*, eds. E. Sheppard, T. Barnes and J. Peck. 226–241. Oxford: Wiley
   Blackwell.

Crang, M., Hughes, A., Gregson, N., Norris, L. and Ahamed, F. 2013. Rethinking governance and value in commodity chains through global recycling networks. *Transactions of the Institute of British Geographers*, 38: 12–24.

Davidson, I. 2012. Automobility, materiality, and Don DeLillo's *Cosmopolis*. *Cultural Geographies*, 19: 469–482.

DeLyser, D. 2014. Tracing absence: Enduring methods, empirical research, and a quest for the first neon sign in America. *Area*, 46(1): 40–49.

DeLyser, D. and Greenstein, P. 2017. The devotions of restoration: Materiality, enthusiasm, and making three "Indian Motocycles" like new. *Annals of the Association of American Geographers*, pp. 1–18. doi:10.1080/24694452.2017.1310020.

DeLyser, D. and Greenstein, P. 2015a. "Follow that car!" Mobilities of enthusiasm in a rare car's restoration. *The Professional Geographer*, 67(2): 255–268.

DeLyser, D. and Greenstein, P. 2015b. Signs of significance: Los Angeles and America's lit-sign landscapes. In *Cities of Light: Two Centuries of Urban Illumination*, eds. S. Isenstadt, M.M. Petty, and D. Neumann. London: Routledge, pp. 101–108.

DeSilvey, C., Bond, S. and Ryan, J.R. 2013. 21 Stories. *Cultural Geographies*, 21(4), 657–672.

Dillon, M. 2002. *Artificial Sunshine: A Social History of Lighting*. London: National Trust.

Edensor, T. 2011. Entangled agencies, material networks and repair in a building assemblage: The mutable stone of St. Ann's Church, Manchester. *Transactions of the Institute of British Geographers*, NS 36: 238–252.

"Fifty Years of Electric Signs" 1956. *Signs of the Times*, May 1956, pp. 17–80.

Foster, R.J. 2006. Tracking globalization: Commodities and value in motion. In *SAGE Handbook of Material Culture*, eds. C. Tilley, W. Keane, S. Kuechler-Fogden, M. Rowlands and P. Spyer. 285–302. London: SAGE Publications.

GrahamS. and Thrift, N. 2007. Out of order: Understanding repair and maintenance. *Theory, Culture and Society*, 24(3): 1–25.

Gregson, N., Crang, M., Ahamed, F., Akhter, N. and Ferdous, R. 2010. Following things of rubbish value: End-of-life ships, "chock-chocky" furniture, and the Bangladeshi middle class consumer. *Geoforum*, 41: 846–854.

Gregson, N. and Crewe, L. 2003. *Second Hand Cultures*. Oxford: Berg.

Gregson, N., Metcalf, A. and Crewe, L. 2007. Moving things along: The conduits and practices of divestment in consumption. *Transactions of the Institute of British Geographers*, N.S. 32: 187–200.

Gregson, N., Metcalfe, A. and Crewe, L. 2009. Practices of object maintenance and repair. *Journal of Consumer Culture*, (9)2: 248–272.

Gregson, N., Watkins, H. and Dalestani, M. 2010. Inextinguishable fires: Demolition and the vital materialisms of asbestos. *Environment and Planning A*, 42(5): 1065–1083.

Gudis, C. 2004. *Buyways: Billboards, Automobiles, and the American Landscape*. London: Routledge.

Jacobs, J. 2006. A geography of big things. *Cultural Geographies*, 13(1): 1–27.

Lepawsky, J. and Billah, M. 2011. Making chains that (un)make things: Waste-value relations and the Bangladeshi rubbish electronics industry. *Geografiska Annaler: Series B Human Geography*, 93(2): 121–138.

Lepawsky, J. and Mather, C. 2011. From the beginnings and endings to boundaries and edges: Rethinking circulation and exchange through electronic waste. *Area*, 43(3): 242–249.

Lowenthal, D. 2016. *The Past is a Foreign Country – Revisited*. Cambridge: Cambridge University Press.

Miller, D. 2011. The power of making. In *The Power of Making*, ed. D. Charny. London: V&A Publishing, pp. 14–27.

Miller, R.C. 2009. *Freeway*. Hermosa Beach, CA: Bombshelter Press.

Pfaff, J. 2010. A mobile phone: Mobility, materiality, and everyday Swahili trading practices. *Cultural Geographies*, 17: 341–357.

Ramsay, N. 2009. Taking-place: Refracted enchantment and the habitual spaces of the tourist souvenir. *Social and Cultural Geography*, 10(2): 197–217.

Rasmussen, C. 1996. Hollywood castles and curious "cures". *The Los Angeles Times*. 18 March. Available at: http://articles.latimes.com/1996-03-18/local/me-48393_1_hol lywood-castles last accessed 2 January 2016.

Ribbat, C. 2013. *Flickering Light: A History of Neon*. London: Reaktion Books.

Rinaldi, T.E. 2013. *New York Neon*. New York: WW Norton & Company, Inc.

Rose, G., Degen, M. and Basdas, B. 2010. More on "big things": Building events and feelings. *Transactions of the Institute of British Geographers*, NS 35: 334–349.

Starr, K. 1999. Landscape electric: A program that renews the city's urban spirit by relighting Philip Marlowe's neon LA. *The Los Angeles Times*. 4 July. Available at: http://articles.latimes.com/1999/jul/04/opinion/op-52802 last accessed 2 January 2016.

Strebel, I. 2011. The living building: Towards a geography of maintenance work. *Social and Cultural Geography*, 12(3): 243–262.

Thomas, N., Harvey, D.C. and Hawkins, H. 2013. Crafting the region: Creative industries and practices of regional space. *Regional Studies*, 47: 75–88.

Treu, M. 2012. *Signs, Streets, and Storefronts: A History of Architecture and Graphics along America's Commercial Corridors*. Baltimore: Johns Hopkins University Press.

Warren, A. and Gibson, C. 2014. *Surfing Places, Surfboard Makers: Craft, Creativity, and Cultural Heritage in Hawai'i, California, and Australia*. Honolulu: University of Hawai'i Press.

14 Geographies of making

Matter, transformation and care

Laura Price and Harriet Hawkins

"The idea that making is its own particular sort of thinking is an appealing one. But it also constitutes a major challenge for anyone who wants to do justice to making through the seemingly inadequate tools of words and ideas" writes Glenn Adamson (2010: 1). Central to Adamson's reflections on the particularity of making are the challenges and tensions posed by thinking and writing of ideas that are embodied and experienced, that are shown, rather than told. How do you put into words, the *feeling of doing*? Throughout this collection our authors have used a range of different methodological skills and techniques to explore the power of making. These have drawn on and developed geographers' cultivated attentiveness to the affectual and emotional qualities of experience, recognising the power of bodies and the more-than-representational. Indeed, one of the essential elements of this volume is the centrality of discussions around "acquired technique, of methods tried and tested, of respected skills and hard-won embodied and material practices" (Hawkins et al., 2015: 222). Yet, as Carr et al. (this volume) suggests, making is 'polyvalent' – which for us, engages not only the intangibility of craft knowledge that poses difficulties for researchers, but rather the ubiquity of making in everyday life such that it may be overlooked and undervalued. Reflecting now, in this short coda, we wish to draw out three cross-cutting things that we feel have emerged from this volume. More specifically, we wish to contemplate the potential of making and the possibilities and transformations we envisage as geographers continue their rich engagements with making, craft, and creativity in theory, research and practice. Of the many possibilities we have selected themes that tie together three pivotal strands from across the whole volume. We begin by considering making as a transformational politics, exploring the scales and spaces through which ideas of transformation and politics are understood. From here we move to reflect on the unequivocal importance of matter and materiality to these discussions. We close by reflecting on the intersection of making and caring, and the importance of care-full making in terms of the geographies of making, but also the geographies that are thus made – whether these are bodily relations between humans and non-humans, or the reimagining of places, communities and environments.

Towards making as a transformational politics

The chapters in this volume have all explored how making transforms our material and social relations, how it produces, makes, repairs and reimagines geographies. These transformations happen across multiple sites, spaces and scales from that of matter and the body, but also occurring at festivals, within supply chains, and across place, communities and environments more broadly. How, though, are we to understand the manifold senses of transformation and politics that evolves? Here we follow Gibson Graham and others, in thinking of transformation, less as grand shifts and changes, and rather more as incremental, daily, perhaps even imperceptible movements in bodies (human and non-humans) attitudes and affective dispositions. Thus, whilst politics might be found within these pages in terms of activism, in reference to institutional structures as well as in terms of policy makers, more often the politics at stake here are micropolitics, that, whilst linked to, are loosed from more standard discernments of the political. From the chapters making emerges as a micropolitical practice that operates through the embodied and material, with micropolitical practices of making operating as ongoing transformational politics. Making, we advocate, is as much about the promise and process of what *can* become, as it is about the product that is made. This echoes Ben Anderson for whom the micropolitical "involves a temporal reorientation of knowledge practices to the emergent and the prospective (what has not-yet become)" (Anderson, 2017: 594). In emotional and affectual terms, we see in the chapters collected here a hopeful politics of transformation, the promise of something as *becoming*: "the hope of micropolitics is that it invites us to learn how to act in the midst of ongoing, unforeclosed situations and experiment with ways of discerning and tending to the 'otherwise'" (Anderson, 2017: 594).

Compellingly, the spaces of many of these transformations are those of the everyday, the ordinary and the vernacular, whilst the practices are as much those of amateurs and those learning to do, as they are of expert or professional makers. Let us start with the small scale, and the oft overlooked practice of knitting, by revisiting Miriam Burke's chapter. These knitters engage in creative processes in ways that have transformed their social and emotional relationships with each other and their understanding of the non-human, the environmental, indeed the elemental too. The social connections that furnish Burke's discussions are similarly co-produced by the hairdresser in Helen Holmes' chapter – as hair is crafted, so too are relationships between the hairdresser and the client: "craft extends not just to the often invisible and intangible skills involved, or to the transient and temporary products and services created, but also to the intimate social relations such encounters can make". Much like objects, matter and material that is handcrafted between skills, bodies and makers – social relations can be simultaneously transient and enduring. In Tim Edensor's chapter we discover the 'essential inclusive-ness' of Slaithwaite's biannual *Moonraking* festival – through production creative participation is manifest as convivial sociality through which places and communities are made. Across these chapters making creates human and non-human encounters that are "mediated, affective, emotive, and sensuous, that (sic) are about animation, joy and fear, and both the opening up and closing down of affective capacity" (Wilson, 2016: 465). Nowhere do we see this more than in Elizabeth Straughan's encounters with skin and flesh during her taxidermy practices, where her relations with non-humans are shaped simultaneously by fascination and disgust. Whilst Burke's and Straughan's chapters might cultivate an attentiveness, even generosity, towards the non-human so too does Richard Ocejo's. This is achieved less through transfor-mational environmental politics and rather through concerns with the com-modification and aestheticisation of (dead) animal bodies for meat. Ocejo's contribution echoes Straughan's in their shared concern for making as an acutely sensitive and embodied practice. A practice that in pursuit of profi-ciency inevitably has to manage tensions, uneven surfaces and material unpre-dictability. Making is a series of material encounters that we might design, and indeed practice, which contain within them an "inherent unpredictability" (Wilson, 2016: 465). As such, these transformations are not guaranteed, and require, as Anderson (2017: 594) says of micropolitics, an attentiveness and openness to the emergent and prospective, rather than the already known. Rebecca Collins articulates this through her phrase 'maker-habitus' which refers to the "acute affordance sensitivity that is an ability to identify the potentialities of materials and material things" – indeed "it is possible to become particularly attuned to alternative forms those things could take".

A commitment to engaging openly and co-productively with the non-human, environmental and elemental is displayed by makers throughout this volume. For Carr et al. this engagement is a geographical imperative, parti-cularly in context of the Anthropocene which "raises questions around what kinds of economies will become necessary and even desirable, in a future characterised by volatile weather events, ecosystem disruptions and resource scarcities". In its essence, exploring the geographies of making seeks to attend to "material geography's need to address its elemental prejudices" particularly as doing so "has the potential to bring a needed radical corrective, not simply to what we study but to how we listen to, question, and make our world" (Jackson and Fannin, 2011: 435). What this volume has shown is how such tendencies are not just the privilege of professional makers and crafters, but increasingly these engagements play out in the space of the vernacular, the everyday, and the amateur. Knott (2015: 83) refers to "resourcefulness, the ability to experiment, management and delegation, the separation of tasks" that "are all rehearsed in amateur space" for, he argues, "an amateur creates highly personal and idiosyncratic spaces that demonstrate particular and unusual relationships to production that nonetheless link back to the eco-nomic and societal reality from which the practice departs" (Knott, 2015: 45).

We must not then, in our embrace of the hopeful and transformatory politics of making, let ourselves become divorced from the economic social and political realities that these practices are situated in. For example, Nicola Thomas's chapter draws our attention to institutions that shape craft aesthetic and taste, referring to how a "very specific vision of craft was enrolled" historically at the Devon Guild of Craftsmen, whilst Ocejo, draws attention to how labels such as "craft-based, handmade, and artisanal" have become highly valued in today's economy. Craftmanship, he argues "holds a central place in the social world of workplaces". Further, as Knott (2015: 83) suggests, the knowledge and skills of a maker enable the amateur to better negotiate the structures of capitalist society. All of this may become refigured again by the latest evolutions in 3D printing which continues the long and complicated relationship of craft, making and technology (Warren, 2014: 2300). Nowhere is the imperative for remaining attuned to the situated contexts and politics of making so clear as in Zoe Collins's chapter on the material politics of sewing harnessed by a range of NGOs. If we might reflect on how institutions, organisations and workplaces discipline makers and their skills, we might equally reflect on the opportunities offered by amateur making practices for sitting within complicated relations with the economic structures and practices of capital. For example, this is clearly illustrated through the research of Luckmann (2015) on Etsy selling, and the interpenetration of capital and the home through the commodification of the home as a site for making.

In short then, making transforms geographies, it recasts connections between humans and non-humans, reimagines new relations between work and leisure, it creates places and communities and attunes our material lives with the vibrancy and needs of our environments. Cautiously, we suggest these experiences are not inherently positive, beneficial and desirable for all involved. It is thus crucial that, as chapters by Collins, Burke and others reiterate, we attend carefully to exactly what kinds of transformations occur through making, through what practices, to whom and with what temporalities and spatialities. Indeed as Black (2017: 707) cautions, "for craft to be used as a tool of substantive social change, it must be engaged with in such a way that it does not become a mask for, and/or an active agent in, processes of injustice, exclusion and privilege".

Towards making material lives

Writing in 2004 Lees-Maffei and Sandino (2004: 214) noted "the allusive significance of materials" in design, art, and craft. We suspect that writing today, the materials involved in making might not be quite so allusive. Across the chapters collected here materials – from yarn, wood and fabric, to hair, skin and flesh – have played a central role in the practices and possibilities of making. As Jane Bennett (2009: 60) argues, "the desire of the craftsperson to see what a metal can *do*, rather than the desire of the scientist to know what a metal *is*, enabled the former to discern a life in metal and thus, eventually, to collaborate more productively with it" (Bennett, 2009: 60). Across this volume, making and makings are addressed in two substantial ways. Firstly, matter is omnipresent as agentive. Secondly, across these chapters practices of making tend to foreground the ongoingness of matter, in other words through practices like reuse, restoration and repair.

That the materials involved in making are agentive is very clear. Sjöholm highlights the material importance of sketchbooks to artists, working together with pencils and paper, the sketchbook solves "practical problems and challenges". Whilst, for Straughan, Ocejo, Burke and Holmes in this volume and to varying degrees, making occurs through agentive materials "acting back" on the human maker and their multiple senses. Whether it be matter that smells, that does not behave as anticipated, that resists the maker's normative experiences of the process, matter we assert does work in the world. Indeed, accounting for material relations in the context of skill, technique, habit and the co-production of humans and non-humans has become a central feature of ethnographies of making. For example, Straughan, reflects on the micro-encounters of taxidermy practices, suggesting that materially focused contributions challenge isomorphic conceptions of the human subject, and assist geographical thinking of the human as emergent with the world. Caitlin DeSilvey and James Ryan, in their chapter on mending, remind us that "the craft of repair always involves, in some sense, the capacity to accommodate the independence of things" and that "the transactions undertaken within repair workshops often have powerful socially integrative effects by fostering shared human appreciation and care for the material qualities and meanings of things". Indeed, across these chapters paying attention to the textures of materials – yarn, hair, skin – or their capacities (to stretch, to curl, to tear) is integral to the book, but these are never materials worked alone. For as Bennett writes, "metal is always metallurgical, always an alloy of the endeavours of many bodies, always something worked on by geological, biological, and often human agencies. And human metalworkers are themselves emergent effects of the vital materiality they work" (2009: 60). As this quotation suggests, an appreciation of materiality and material practices of making cannot be seen apart from the production of social life and human experience concerned with those practices. We see this in Dydia DeLyser's discussions of the restoration of neon lighting. Here, the working of glass, light and metal becomes a set of skills and dispositions cultivated within the maker's bodies, habits, knowledge, and practices that shape muscles and future abilities.

If stories of materials and making can be micro-stories of the skill of manipulating tools and matter, then they can also be thoughtful narrations of mobilities and immobilities, and of ongoing material lives. Consequently, the chapters written by DeLyser, Holmes and DeSilvey and Ryan detail the role of making in extending and evolving the material lives of objects through practices of restoration and repair. We may conclude, or infer then, that making is not just about tracing through original moments of production, but rather about extending and evolving discussions of the material lives of objects which take into account their ongoingness, as they are patched up, repurposed and otherwise reused, in place of being discarded. Such attentiveness to material lives brings with it, perhaps unsurprisingly, both explicit and implicit resistance to contemporary modes of consumption and discard and thus inherent possibilities as Collins (this volume) explores for the cultivation of sustainability and environmentalisms.

Thus, making opens up myriad opportunities for telling stories of labour and capital otherwise. We would argue that engaging with making, can and should frame producers as part of global commodity and trade circuits, but it can also open up the possibilities to share maker-stories and voices, identities and 'pleasurable' making experiences – so that materials, labour processes, and production methods can be envisaged otherwise. This is not of course to overlook the challenges that Zoe Collins's chapter raises, where stories of the "taking advantage of the assumed benignity of sewing apparatus" are powerful and evocative – here "sewing is not only a tool for the creation of textile items, but also for activism, discipline, the dissemination of ideology, and the generation of income". It is to recognise the possibilities for other stories of labour and capital to be told. Whilst we are aware of the relative focus of our collection in the Global North, since its evolution a series of other accounts have begun to enrichen discussions within wider geographies of making. Daya (2016: 133) draws attention to crafting stories in "informal and market spaces of the city (Cape Town) that are so characteristic of many cities in the global South" – here "the pleasure that craft producers took in consumers' appreciation of their work was not trivial. It amounted to more than simply a valuing of the product; rather, it helped to generate their sense of themselves as artists". Similarly, Klocker et al. (2017: 13) tell the story of footballs produced from wasted plastic bags in the Majority (developing) world; makers find opportunities to think differently with materials that have arisen in unexpected places; it is argued, "contra the prevailing rhetoric" – "young men who make and play with plastic-bag footballs do not feel deprived. They feel joy, and care for their bespoke-made objects. They do not need pity or saving" (Klocker et al., 2017: 13). The pleasure of hand-making is a fundamentally human experience, and the telling of making stories and creativity – of conviviality, generosity of time and skill, and expertise – should not be confined to cities in the Global North, but to vernacular spaces, informal markets, and developing countries so that "we see craft inhabiting the secret folds of everyday life, the domestic, the emblematic, and the functional with the stories of nationhood, power, resistance, living and lineage" (Ravetz et al., 2015: 189). We might also consider the politics of *making-do*, or the *make-shift* and the how this linguistic accruement reflects geographies of potential precarity, austerity, and the ad-hoc shaped from necessity. For example Thieme (2017: 15) refers to "'hustling' as a way to foreground the everyday practices of makeshift urbanism through which (young) lives on edge produce and are produced through these uncertainties". If we value creativity, human agency to make, create, and survive, there is a challenge of not letting such a valuation translate into the perpetuation of "the ordinary and makeshift urban practices that occur across geographies (and increasingly in the Global North), where 'crises' become unexceptional, and where coping with uncertainty is normalized" (Thieme, 2017: 2). These challenges and possibilities lead us to our final point around making and practices of care.

Towards care-full making

The practices, processes and geographies of making explored in this volume broadly and variously articulate forms of care – towards materials, things, objects, people, places and environments. In many ways the hand-made has become shorthand for commodities that are potentially more caring; this is reflected by the investment of time, processes of personalisation, and a consideration of who, what, where, and why that is perhaps *neglected* in other forms of consumption. We have thought (carefully!) about our view on this and in this final section of the conclusion we want to reflect on some of the complexities of making and caring, and explore the possibilities and pitfalls of a care-full sense of the geographies of making.

The framing of making as a practice of care is historically derived from the co-production of feminised care practices and arts associated with home making (Hackney 2006; Luckman, 2013). Fundamentally, as a form of creativity, embodied practices connoted as craft have been "pushed to the margins of intellectual life" (Dormer, 1997: IX) and have often lacked "detailed historical analysis" (Greenhalgh, 1997: 19). Moreover, spatially, "we see craft inhabiting the secret folds of everyday life, the domestic, the emblematic, and the functional" (Kettle et al., 2015: 189). As crafts are brought out of the closet, fabric stashes released out of the attic, humble sheds paid attention to as sites of lively creativity – it has been easy for geographers to enthusiastically embrace these stories with supportive, critical, but often celebratory fashions. This is complicated further through the complexities of creativity and that certain forms of making have, historically, been gendered and thus neglected and cast as women's work (or leisure). How, we ask, can we continue to share stories of making as practices of care without making moral judgements? As Meah and Jackson (2017: 2069) have argued in the context of food: 'convenience' and 'homemade' foods should not be regarded as mutually exclusive categories, with the "latter perceived as inherently more indicative of care than the former, but should instead be understood in terms of the values which they are subjectively intended to achieve". We feel this deserves more attention, particularly whilst engaging with making practices that seek to subvert the capitalist and consumer economy. To conclude we want to identify three interlinked senses of making and care, but we also wish to issue a caveat.

Firstly, there is care understood in terms of precision, as making comes to be associated with technique and skills and concern for materials and material lives, that as we have suggested above can cultivate a particular kind of attentiveness. The place of care within our research practice and methodologies should not be disregarded. Whether this be taking care to learn crafts and skills ourselves, or continuing to work with makers, tinkerers, crafters and menders who may at first display incredulity at sociological and geographical interest in their skills, hobbies and interests. Secondly, there is the way that care is expressed through objects and projects that are produced whether it be things that are made, that express care for people beyond the object or gift shared, or whether it be the making of a home. Finally, as the chapters in this volume have explored, making practices cultivate care for and within communities that are human and non-human. Working together with people on shared projects, is not "merely pleasant-but-optional 'icing on the cake' of individuals' lives, but is absolutely essential for personal well-being and for a healthy, secure, trust-worthy society" David Gauntlett (2011; 162). So care cultivated through making might be about care for those we make, collaborate and co-produce with, but might also be about a shifting in affective relations. As a number of our authors here have expressed, it is hoped that this cultivation of care might result in potentially more ethical consumption and environmentally minded behaviours.

Before though, we get carried away with care-full understandings of making we need to temper this with a careful attention to the challenges of making. So, as Collins explored here, making is not always an appropriate way to express 'care' for distant others. As Richard Sennett (2012) reminds us, practical and material based skills do not always translate as being better equipped with social skills through which to live life. Therefore, we issue a caution, to remain critical and attentive to the wider geographies that *produce* making and also to document the less enchanting experiences of making and doing: the frustrations, failures, abuses, pain and generally uncomfortable geographies of making that are part of learning to do. Whilst immersion, flow experience, and project work is rich and sustaining – let us think about the potential to be *lost* in our work. This is particularly critical when approaching making through the lens of care, as challenging the macropolitics of production and consumption through the micropolitical can be laborious and demanding – let us remember that "it is clear that consumers care about these issues, even if they lack the capacity to enact their perceived responsibilities" (Meah and Jackson, 2017: 2069). In short, let us not neglect the wider cultural, social and economic practices and contexts that embodied making practices sit within.

With those caveats, we still think that care is pivotal in helping us to appreciate how to attend to the geographies of making in ways that also attend how we make geographies. Care comes together with making in myriad ways, attuning us to practices of those we study and ourselves, engaging with complex questions of gender relations, fluid boundaries between home and work, between capital and creativity, between economic and leisure. Care also calls us towards an appreciation of how making cultivates ethical relations between humans, and between humans and non-humans (living and non-living). Indeed, thinking about care and making offers a bridge between the different perspectives that this book sought to bring together, perspectives that attend to the rich tapestries of research into the geographies of making, weaving concerns with the spaces, practices and materials through which making happens, with the complex methods and means by which making re-makes, repairs and reimagines our geographies.

References

Adamson, G. (2010) *The Craft Reader*. London: Berg.

Anderson, B. (2017) Hope and micropolitics. *Environment and Planning D: Society and Space*, 35(4), 593–595.

Bennett, J. (2009) *Vibrant Matter: A Political Ecology of Things*. Duke University Press.

Black, S. (2017) KNIT+ RESIST: Placing the Pussyhat Project in the context of craft activism. *Gender, Place & Culture* (Online first), 1–15.

Carr, C. and Gibson, C. (2016) Geographies of making: Rethinking materials and skills for volatile futures. *Progress in Human Geography*, 40(3), 297–315.

Daya, S. (2016) Ordinary ethics and craft consumption: A Southern perspective. *Geoforum*, 74, 128–135.

Dormer, P. (ed.) (1997) *The Culture of Craft*. Manchester: Manchester University Press.

Gauntlett, D. (2011) *Making is Connecting: The Social Meaning of Creativity from DIY and Knitting to YouTube and Web 2.0*. London: Polity Press.

Greenhalgh, P. (1997) The history of craft. In Dormer, P. (ed.) *The Culture of Craft*. Manchester: Manchester University Press.

Hackney, F. (2006) Using your hands for happiness: Home craft and make do and mend in British women's magazines in the 1920s and 1930s, *Journal of Design History*, 19(1), 23–38.

Hawkins, H., Cabeen, L., Callard, F., Castree, N., Daniels, S., DeLyser, D., Neely, H.M. and Mitchell, P. (2015) What might GeoHumanities do? Possibilities, practices, publics, and politics. *GeoHumanities*, 1(2), 211–232.

Jackson, M. and Fannin, M. (2011) Letting geography fall where it may – aerographies address the elemental. *Environment and Planning D: Society and Space*, 29, 435–444

Klocker, N., Mbenna, P. and Gibson, C. (2017) From troublesome materials to fluid technologies: making and playing with plastic-bag footballs. *Cultural Geographies* [Online First].

Knott, S. (2015) *Amateur Craft: History and Theory*. London: Bloomsbury Publishing.

Lees-Maffei, G. and Sandino, L. (2004). Dangerous liaisons: Relationships between design, craft and art. *Journal of Design History*, 17(3), 207–219.

Luckman, S. (2013) The aura of the analogue in a digital age: Women's crafts, creative markets and home-based labour after Etsy. *Cultural Studies Review*, 19, 249–270.

Luckman, S. (2015) *Craft and the Creative Economy*. London: Palgrave Macmillan.

Sennett, R. (2012) *Together: The Rituals, Pleasures and Politics of Co-operation*. London: Penguin.

Meah, A. and Jackson, P. (2017) Convenience as care: Culinary antinomies in practice. *Environment and Planning A*, 49(9), 2065–2208.

Ravetz, A., Kettle, A. and Felecy, H. (2013) *Collaboration through Craft*. London: Bloomsbury.

Thieme, T.A. (2017) The hustle economy: Informality, uncertainty and the geographies of getting by. *Progress in Human Geography* [Online First].

Warren, A. (2014) The agency and employment experiences of non-unionized workers in the surfboard industry. *Environment and Planning A*, 46(10): 2300–2316.

Wilson, H.F. (2016) On geography and encounter: Bodies, borders, and difference. *Progress in Human Geography*, 41(4), 451–471.

Index

Printed in the United States
by Baker & Taylor Publisher Services